RAND NATIONAL DEFENSE RESEARCH INSTITUTE

ISSUES WITH

Access to Acquisition Data and Information

IN THE DEPARTMENT OF DEFENSE

Policy and Practice

Jessie Riposo, Megan McKernan,
Jeffrey A. Drezner, Geoffrey McGovern,
Daniel Tremblay, Jason Kumar,
Jerry M. Sollinger

Prepared for the Office of the Secretary of Defense

For more information on this publication, visit www.rand.org/t/RR880

Library of Congress Cataloging-in-Publication Data is available for this publication.
ISBN: 978-0-8330-8892-5

Published by the RAND Corporation, Santa Monica, Calif.

Preface

Acquisition data underpin the management and oversight of the U.S. defense acquisition portfolio. However, balancing security and transparency has been an ongoing challenge. Some acquisition professionals are not getting the data they need to perform their assigned duties or are not getting the data and information in an efficient manner. To help guide the Office of the Secretary of Defense (OSD) in addressing these problems, the RAND Corporation identified access problems at the OSD level—including those organizations that require access to data and information to support OSD, such as analytic support federally funded research and development centers and direct support contractors—and evaluated the role of policy in determining access. The study also involved a limited review of how data are shared between OSD and the military departments.

This report should be of interest to government acquisition professionals, oversight organizations, and, especially, the analytic community.

This research was sponsored by the Office of the Secretary of Defense and conducted within the Acquisition and Technology Policy Center of the RAND National Defense Research Institute, a federally funded research and development center sponsored by the Office of the Secretary of Defense, the Joint Staff, the Unified Combatant Commands, the Navy, the Marine Corps, the defense agencies, and the defense Intelligence Community.

For more information on the RAND Acquisition and Technology Policy Center, see http://www.rand.org/nsrd/ndri/centers/atp.html or contact the director (contact information is provided on the web page).

Contents

APPENDIXES

Figures and Tables

Figures

Tables

Summary

Acquiring military equipment is big business. The value of the current portfolio of major weapon systems is about $1.5 trillion. Managing such a huge portfolio requires access to an enormous amount of acquisition data, including the cost and schedule of weapon systems (both procurement and operations), information about how they perform technically, contracts and contractor performance, and program decision memoranda. The Office of the Under Secretary of Defense for Acquisition, Technology and Logistics (OUSD[AT&L]) and those working for it need access to such data to track acquisition program and system performance and ensure that progress is being made toward such institutional goals as achieving efficiency in defense acquisition and delivering weapon systems to the field on time and on budget.

A range of organizations need access to the data for different purposes (e.g., management, oversight, analysis, administrative). Such organizations include various offices of the Department of Defense (DoD), federally funded research and development centers, university-affiliated research centers, and a host of support contractors.

But getting access to the data to carry out the analyses requested by DoD is not always easy, or in some cases even possible. At times, the data carry dissemination restrictions that limit their distribution. In other cases, proprietary information is the property of a commercial firm and may not be released without that firm's explicit permission. The Office of the Secretary of Defense (OSD) asked the RAND National Defense Research Institute to identify the problems and challenges associated with sharing unclassified information and to investigate the role of policies and practices associated with such sharing. This report details the issues associated with gaining access to what is called *Controlled Unclassified Information* (CUI).

What We Found

The process for gaining access to data is inefficient and may not provide access to the best data to support analysis. Government personnel and those supporting the government sometimes do not get their first choice of data, and even that data may take a long time to receive. They may be forced to use alternative sources, which often have data of lower quality, which might be dated and thus less accurate, or be subject to a number of caveats. While the consequences of these limitations are undocumented and difficult to assess and quantify, the results of these analyses can be inferior, incomplete, or misleading.

Two groups of people face particular challenges in gaining access to data: OSD analytic groups and support contractors. OSD analytic groups often do not have access to the originators of the data, which precludes them from going to the primary source. They also tend to have poor visibility of all viable data sources, which encourages inefficient data-seeking practices. Direct support contractors have problems similar to OSD analysts, but these problems can be compounded by laws, regulations, and policy that restrict access to certain types of information (especially nontechnical proprietary data that originate and are labeled outside the government), which introduces extreme inefficiencies. Support contractors require special permissions to view nontechnical proprietary data.

Difficulty in gaining access occurs for several reasons.

- Data access policy is highly decentralized, not well known, and subject to a wide range of interpretation.
- The markings for unclassified information play a significant role in access. The owner or creator of a document determines what protections or markings are required. However, marking criteria are not always clear or consistently applied. In fact, management and handling procedures for many commonly used markings are not clearly described anywhere. Once marked, getting the labels changed can be difficult. When information is not marked, the burden of handling decisions is placed on the receiver of the information.
- Institutional and cultural barriers inhibit sharing. The stove-piped structure of DoD limits visibility and sharing of data and information. Institutional structure and bureaucratic incentives to restrict data access are exacerbated by policy and guidance to protect information. The result is a strong conservative bias in labeling and a reluctance to share. A lack of trust and established relationships can hinder sharing.

Options for Improving Data Sharing

The variety of identified problems may be addressed in many ways. Each potential option requires further analysis and investigation. We offer initial thoughts to deal with the issue of access to proprietary data, as well as the general confusion regarding policy.

Options to Address Problem of Proprietary Data Access

There are several potential options to resolve the problem of access to proprietary data.

- The Under Secretary of Defense for Acquisition, Technology and Logistics (USD[AT&L]) could seek additional billets and insource any functions that require access to proprietary data. However, this would require Office of Professional Management and congressional support.
- USD(AT&L) could seek relief through a reallocation of billets to functions that currently require access to proprietary information. This would require cross-organizational prioritization, a difficult process.
- General access could be established for all direct support contractors. This would require legislative or contractual changes. Current legislation, Title 10 U.S. Code, Section 129d, allows litigation support contractors to view proprietary information. Similar legislation might be pursued for all support contractors.
- Alternatively, additional contractual language could be placed on all DoD acquisition contracts granting support contractors restricted access to their data. The direct support contractors who receive the data would have to demonstrate company firewalls, training, personal agreements, and need to know akin to those for classified information.

- The government could seek an alternative ruling on the nondisclosure requirements, whereby blanket nondisclosure agreements could be signed between the government and a direct support organization, or a company and a direct support organization to cover multiple tasks.

Each of these options would require further analysis and coordination with Office of the General Counsel and Defense Procurement and Acquisition Policy (and Congress in the first and third options).

Options to Address Policy Confusion

There are also several options to addressing the confusion regarding policy.

- OUSD(AT&L) could create and maintain a central, authoritative online resource that references all relevant guidance on information management, handling, access, and release for acquisition data. This would require identifying the relevant policy and posting new policies as they become available.
- However, an online resource may not address the issue of the workforce having a general lack of expertise and insight regarding the existing policy and guidance. To cope with this problem, OUSD(AT&L) could also consider providing additional training for its staff on the identification and protection of data. This could be an annual online training for all OUSD(AT&L) staff and contractors.
- In areas where conflicting interpretations of guidance are particularly problematic, such as with For Official Use Only (FOUO) and proprietary information, additional guidance about how to determine whether information is FOUO or proprietary in the first place would be helpful. The guidance should provide specific examples of information that is considered protected, guidelines for determining whether specific information qualifies, and details regarding handling procedures for this information, to include access privileges.
- Directives and incentives could be established so that markings that appear to be incorrect are challenged and not taken only on a company or individual's claim. If more-detailed determination guidance is available, it could be used to assess the validity of a marking. A process should be in place for challenging markings, and it should be exercised.

There are important reasons for restricting access that require balancing control with granting more access. In information assurance and security policy, there is an understanding that no individual should have unfettered access to all data. Given the inherent complexity in securing data and sharing data, any solutions to problems associated with data sharing must be well thought out to avoid the multitude of unintended consequences that could arise.

Acknowledgments

We would like to thank the sponsors of this study: Mark Krzysko, deputy director, Enterprise Information, Office of the Under Secretary of Defense for Acquisition, Technology and Logistics, Acquisition Resources and Analysis (OUSD[AT&L]/ARA) Directorate; Steven Miller, director, Advanced Systems Cost Analysis Division, Office of the Secretary of Defense (OSD), Cost Assessment and Program Evaluation (CAPE); and Gordon Kranz, deputy director, Earned Value Management, OUSD(AT&L)/Performance Assessments and Root Cause Analysis. We would also like to thank our project monitor, Jeff Tucker, acquisition visibility capability manager, OUSD(AT&L)/ARA, for his guidance and support throughout this study. Also in OSD, we thank the following people who provided us with additional background information that informed our analysis: Douglas Shontz; Chad Ohlandt; Larrie Ferreiro, director of research, Defense Acquisition University; Rob Flowe, OSD Studies and Federally Funded Research and Development Center Management, OUSD(AT&L)/ARA; and Bess Dopkeen, Advanced Systems Cost Analysis Division, OSD CAPE. We'd also like to thank the Acquisition Visibility team and everyone else who volunteered their valuable time to describe their points of view to the RAND study team. We are also grateful to the contractors Brian Baney and Melissa Tengs, who helped facilitate communication with the Office of Enterprise Information in OUSD(AT&L)/ARA.

At RAND, Irv Blickstein and Martin Libicki helped us consider the problems identified in this study. We are very grateful to the two formal reviewers of this document, who helped improve it through their thorough reviews: Ed Keating and Philip Antón. We also thank Melissa Bradley for her help formulating the interview protocol and Christine Dozier and Maria Falvo for their administrative support.

Finally, we would like to thank the director of the RAND Acquisition and Technology Policy Center, Cynthia Cook, and the associate directors, Marc Robbins and Paul DeLuca, for their insightful comments on this research.

CHAPTER ONE

Introduction

Acquisition data are vast and include such information as the cost of weapon systems (both procurement and operations), technical performance, contracts and contractor performance, and program decision memoranda. These data are critical to the management and oversight of the $1.5 trillion portfolio of major weapon programs by the Office of the Under Secretary of Defense for Acquisition, Technology and Logistics (OUSD[AT&L]).[1] Data collection and analysis enable the Department of Defense (DoD) to track acquisition program and system performance and ensure that progress is being made toward such institutional goals as achieving efficiency in defense acquisition and delivering weapon systems to the field on time and on budget.

Many organizations or groups need access to this information for a variety of purposes (e.g., management, oversight, analysis, administrative). These organizations include various offices of DoD, federally funded research and development centers (FFRDCs), university-affiliated research centers (UARCs), and a range of support contractors. For example, an FFRDC may need cost and schedule information to determine whether a weapon system was delivered on time and within budget. Or a support contractor may be responsible for managing a centralized information system for DoD that contains information about specific procurement programs. Note that that situation does not include classified data, which is not a topic of this report.[2]

However, these organizations may have difficulty getting access to these data. Some examples of the types of issues identified by individuals within DoD offices include:

> It took me three months, multiple e-mails and phone calls, to get a one-hour meeting with five SES [DoD senior executive service–level employees] to view data that *might* be proprietary.

> Each access account I create is like five touch points between an email, phone call, their POC, certificate handling, vetting. It's a lot of work.

[1] U.S. Government Accountability Office, *Defense Acquisitions: Assessments of Selected Weapon Programs*, Washington, D.C., GAO-14-340SP, March 2014, p. 3.

[2] Classified information is any information designated by the U.S. government for restricted dissemination or distribution. Information so designated falls into various categories depending on the degree of harm its unauthorized release may cause. This report does not deal with classified information.

> If there are dozens of support contractors and dozens of prime contractors and I have to get an NDA [nondisclosure agreement] for each support contractor and prime contractor combination, it's a lot of work.

Examples of the types of issues identified by FFRDC, UARC, and direct support contractors include:

> The sponsor has to have access, then request a download of several documents I need, then transfer the data to me.

> I couldn't get access because I didn't have a .mil e-mail address.

In some cases the information may be the intellectual property of a commercial firm. Sometimes such information is designated *proprietary*. This information requires the permission of the firm that owns the information to use it. The process of getting permission to use the information can be time-consuming, may never yield permission, or is simply too onerous. An example of the third possibility is a database that has proprietary information from many firms, requiring support contractors to sign NDAs with each firm, which could number many dozens and take a very long time.

The Office of the Secretary of Defense (OSD) asked the RAND National Defense Research Institute to identify the problems and challenges associated with sharing unclassified information and to investigate the role of policies and practices with such sharing.

Approach

We pursued a three-pronged approach with the objective of defining and evaluating any data-sharing problems. The first part of the approach was a policy review. We began by reviewing DoD directives, instructions, manuals, and guides, along with executive orders, legislation, and regulations concerning information management. The objective of the review was to develop a framework for understanding what governs information sharing in DoD acquisition. As part of this search, we also looked at a limited number of key federal policies that might affect data sharing within DoD. We presented a list of relevant policies in Appendix A.

We then met with individuals within OSD to discuss information sharing, which is the second part of our approach. The discussion topics can be found in Appendix B. We used these discussions to help identify information-sharing practices and issues associated with data access and releasability. The discussions also helped us identify relevant policies and practices. We selected a sample of offices within OUSD(AT&L) to reflect a variety of roles in the acquisition process. We spoke with data owners, maintainers, users, and individuals involved with the governance of information. We categorized the offices represented in the sample by their missions and roles. This step led to three main categories of OSD offices:

- functional and subject-matter experts
- Overarching Integrated Project Team/Defense Acquisition Board (OIPT/DAB) review offices
- analysis offices.

Within OSD, the functional and subject-matter experts mainly work within a specialty (e.g., testing, cost, systems engineering, contracts, earned value). Those in the OIPT offices are primarily responsible for direct interaction with acquisition programs to review portfolio status and program readiness as programs move through the acquisition process. The analysis offices conduct a variety of crosscutting analyses in defense acquisition. The offices that fall into these categories appear in Table 1.1. We also interviewed service-level acquisition personnel to determine the role that the services play in DoD data sharing.

Finally, we asked about several major central acquisition data repositories. The intent was to understand the benefits of and difficulties in using central repositories as opposed to direct interaction with program offices. Appendix C provides more-detailed information on selected central repositories: Acquisition Information Repository (AIR), Defense Acquisition Management Information Retrieval (DAMIR), Defense Automated Cost Information Management System (DACIMS), Defense Technical Information Center (DTIC), Federal Procurement Data System–Next Generation (FPDS-NG), PARCA's EVM repository, and OUSD(AT&L) Workforce Data Mart.

Our goal for the interviews was to collect the following information regarding interviewees' data sharing and practices:

- role in the acquisition process
- data needed to perform one's job
- how data are handled, obtained, and provided to others
- data access or release problems
- data-sharing recommendations.

The questions that we used as a basis for our interviews appear in Appendix B.

The final part of our three-pronged approach involved conducting two case studies to illuminate key issues and challenges associated with data access. Both reflect (or embody) the perception of several key data access issues. The first case study examines the use of propri-

Table 1.1
Offices with Roles in the Acquisition Process

Office Category	Offices
Functional and subject-matter experts	• OUSD(AT&L) Performance Assessments and Root Cause Analyses (PARCA) Earned Value Management (EVM) • OSD Cost Assessment and Program Evaluation (CAPE) • OUSD(AT&L) Human Capital Initiative (HCI) • OUSD(AT&L) Defense Procurement and Acquisition Policy (DPAP) • OUSD(AT&L) Developmental Test and Evaluation (DT) • OUSD(AT&L) Systems Engineering (SE)
OIPT/DAB review offices	• OUSD(AT&L) Deputy Assistant Secretary of Defense (DASD) Tactical Warfare Systems (TWS) • OUSD(AT&L) DASD Space, Strategic and Intelligence Systems (SSI) • OUSD(AT&L) DASD Command, Control, Communication, Cyber and Business Systems (C3CB)
Analysis offices	• OUSD(AT&L) Acquisition Resources and Analysis (ARA) • OUSD(AT&L) Defense Acquisition University (DAU) • DPAP • FFRDCs • OUSD(AT&L) PARCA (outside EVM)

etary information (PROPIN) in acquisition, with a particular focus on earned value data. The second looks at the various central data repositories that OSD maintains and uses. More specifically, the focus was on the background, benefits, and problems associated with these repositories. During our introductory interviews, we heard about problems with using, managing, and accessing PROPIN due to the need to involve direct support contractors in the collection and analysis of these data. Such relationships require the use of NDAs to help prime contractors and subcontractors protect their information. Both case studies are informed by the interview results and policy analysis.

Organization of This Report

The remainder of this report is organized as follows. Chapter Two charts the policy landscape for information sharing and what shapes it. Chapter Three describes the issues surrounding access to and release, management, and handling of acquisition data and related information. The discussion reflects perceptions of the issues and challenges in this area and the factors that precipitate those challenges. Chapter Four presents a case study on proprietary data, which allows a more detailed examination of a particularly problematic area. Chapter Five integrates the results of our policy review and interviews to define more precisely prevailing data access problems, the reasons for those problems, and the context within which they will need to be resolved. The chapter also provides options for improving data sharing. The report concludes with four appendixes. Appendix A lists relevant data-sharing policies collected and considered in our analysis. Appendix B presents a summary of the discussions we had with OSD stakeholders. Appendix C is a case study on selected OSD central repositories, which capture and store acquisition data and make them available for various uses. Finally, Appendix D offers a brief overview of technical data rights.

CHAPTER TWO

The Policy Landscape

The policy landscape governing information sharing is vast and decentralized. Feedback from numerous OSD employees indicated that this decentralization has made it more difficult for individuals to locate information or guidance of a particular nature, or to identify the organization responsible for providing such guidance. This chapter describes the disjointed nature of setting policy and then discusses the document markings that guide the release of information.

Decentralization

Offices that promulgate these policies are shown in Figure 2.1. The policies cover a variety of topics. Nearly half of the identified policies describe procedures for handling a specific

Figure 2.1
DoD Offices Issuing Data-Management, Access, Release, and Handling Policies

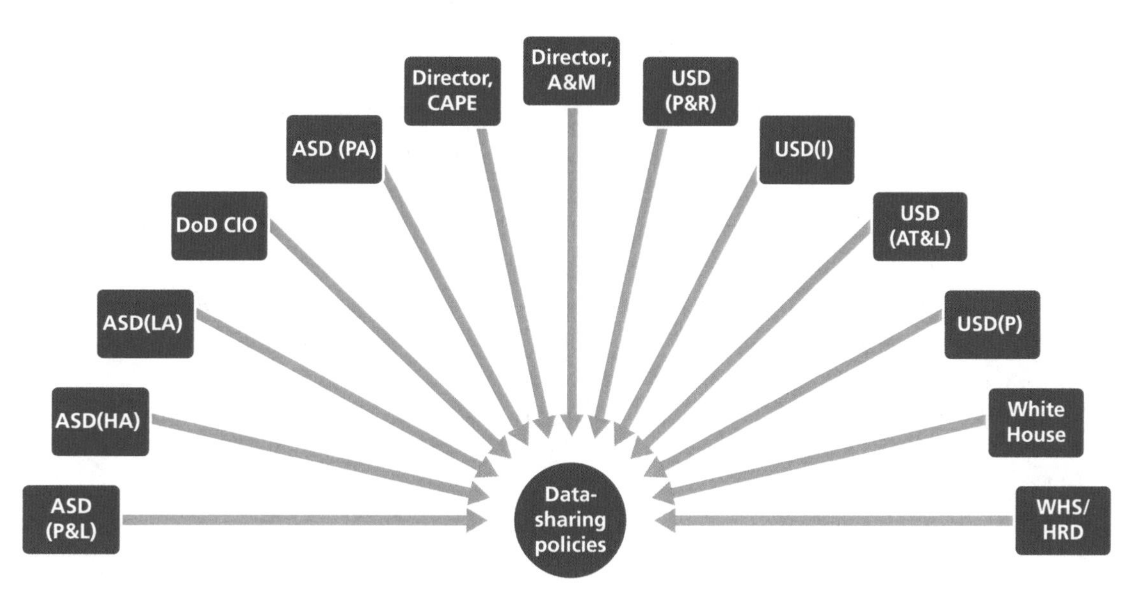

NOTE: ASD(P&L) = Assistant Secretary of Defense for Production and Logistics; ASD(HA) = Assistant Secretary of Defense for Health Affairs; ASD(LA) = Assistant Secretary of Defense for Legislative Affairs; DoD CIO = DoD chief information officer; ASD(PA) = Assistant Secretary of Defense for Public Affairs; A&M = Administration and Management; USD(P&R) = Under Secretary of Defense for Personnel and Readiness; USD(I) = Under Secretary of Defense for Intelligence; USD(AT&L) = Under Secretary of Defense for Acquisition, Technology and Logistics; USD(P) = Under Secretary of Defense for Policy; WHS/HRD = Washington Headquarters Services, Human Resources Directorate.

RAND *RR880-2.1*

Table 2.1
Interviewee Questions Regarding Policy

Interviewees' general inquiries regarding policy and data sharing indicate clarification is needed in multiple areas	More-specific inquiries came from interviewees regarding PROPIN, FOUO, and contractor roles
What constitutes legitimate rationale for gaining access to data?	What can be considered PROPIN?
Who is responsible for removing the caveats when something is no longer source sensitive or classified?	Who can determine if something is PROPIN?
Who can correct a label on a document that is clearly wrong?	What is the policy for releasing PROPIN?
What determines "need to know"?	What constitutes FOUO?
What determines "government only"?	Is there guidance on FOUO?
Do we have a policy that access to data should be written into all contracts?	How can FOUO be remarked?
Is there policy/guidance which dictates where information can flow?	Can FFRDCs and UARCs be considered direct support/ direct report?
	Is there a policy for granting FFRDCs special access permissions?
	Is there any clarifying guidance on how to work with FFRDCs?

type of information, such as technical data or personnel information. In a few instances, the instruction describes a specific type of information but does not detail a specific procedure. A small number of issuances describe the release of information to specific groups of people—for example, the release of information to Congress or to the public, or procedures for sharing information with multinational partners or the Secret Service. Some policies describe various aspects of the information-sharing environment related to accessing information technology (IT), securing IT, or transmitting information and data using information systems or technology. The rest of the issuances describe procedures, programs, or policies more generally. Appendix A lists the policies reviewed in this study, the organization responsible for each policy, the subject of the policy, and the applicability of the policy.

Perhaps as a result of the decentralization and other factors, the individuals we spoke with did not have a good understanding of existing policy. Table 2.1 shows a sampling of questions that were asked of the study team.

These questions demonstrated the confusion surrounding the existing guidance, and illuminated the challenges staff have with determining what is and is not proprietary and For Official Use Only (FOUO).

Marking Criteria

Even though a number of offices are involved in developing policy, the policy for managing and handling unclassified information is incomplete. Unclassified information, the access to which requires special controls, is referred to as Controlled Unclassified Information (CUI). To facilitate the sharing of this type of information, President Barack Obama issued Executive

Order (EO) 13556 on CUI.[1] The order established a program for safeguarding information that is not classified but requires safeguarding. The order states,

> At present, executive departments and agencies (agencies) employ ad hoc, agency-specific policies, procedures, and markings to safeguard and control this information, such as information that involves privacy, security, proprietary business interests, and law enforcement investigations. This inefficient, confusing patchwork has resulted in inconsistent marking and safeguarding of documents, led to unclear or unnecessarily restrictive dissemination policies, and created impediments to authorized information sharing. The fact that these agency-specific policies are often hidden from public view has only aggravated these issues.

However, at the time of this writing, the specific categories and subcategories of CUI were still being developed. In the meantime, current procedures for "sensitive but unclassified (SBU) information remain until CUI is implemented. CUI categories may not be used until the phased implementation for marking is set, and markings are approved and published in the CUI Registry." The date for implementation is yet to be determined. This is problematic because SBU is also vague and inconsistently applied. The publication *What Is CUI?* states,

> There are currently over 100 different ways of characterizing SBU information. Additionally, there is no common definition, and no common protocols describing under what circumstances a document should be marked, under what circumstances a document should no longer be considered SBU, and what procedures should be followed for properly safeguarding or disseminating SBU information. As a result of this lack of clarity concerning SBU, information is inconsistently marked. This puts some information at risk through inadequate safeguarding and needlessly restricts other information by creating impediments. CUI reform is designed to address these deficiencies, in that it will provide a common definition and standardize processes and procedure.[2]

Within the SBU category, interviewees identified two markings as particularly problematic: FOUO and PROPIN.[3] The issues and challenges associated with proprietary data are described in Chapter Four. Despite available guidelines, FOUO marking criteria are not always clear or consistently applied, which can lead to incorrect markings or conservative approaches to marking, which hinder or prevent sharing.

For Official Use Only

FOUO is a control marking for unclassified information that, upon review as a result of a request for information, may be determined to be information that can be withheld from the public if disclosure would reasonably be expected to cause a foreseeable harm to an interest

1 Executive Order 13556, *Controlled Unclassified Information*, Washington, D.C.: The White House, November 4, 2010.

2 Controlled Unclassified Information Office, *What Is CIU? Answers to the Most Frequently Asked Questions*, Washington, D.C.: U.S. National Archives and Records Administration, 2011, p. 3.

3 A distribution marking governs the ability of the owner of the information to share it with someone else. Distribution markings can be applied to classified and unclassified information. In our policy review, several terms were used to describe these concepts: *distribution marking*, *distribution statement*, *dissemination control marking*, *restrictive markings*, and *classification markings*, among others. The research team was not able to identify a single definitive source that outlined and described all of the possible markings for information.

protected under the Freedom of Information Act (FOIA).[4] The FOIA grants individuals the right to federal agency information unless records are protected by an exemption or exclusion. Exemptions and exclusions are presented in Table 2.2.

Exemptions 1 through 6 are most applicable to DoD. Exemptions 2 through 6 are most applicable to DoD CUI. Implementation of the FOIA has required clarification over the years, as evidenced by the numerous amendments that have been made to the act since its inception.[5] In September 2009, the Office of Government Information Services was opened to offer mediation services as an alternative to litigation for resolving FOIA disputes. To help with the interpretation and determination of FOIA exemptions, the U.S. Department of Justice has the Office of Information Policy that can be consulted.

Requirements for and a description of FOUO information appear in DoD Manual 5200.01, Volume 4.[6] While information designated FOUO may be generally disseminated

Table 2.2
Freedom of Information Act Exemptions

Exemption	Exemption Description
Exemption 1	Information that is classified to protect national security. The material must be properly classified under an executive order.
Exemption 2	Information related solely to the internal personnel rules and practices of an agency.
Exemption 3	Information that is prohibited from disclosure by another federal law. Additional resources on the use of exemption 3 can be found on the U.S. Department of Justice "FOIA Resources" web page.
Exemption 4	Information that concerns business trade secrets or other confidential commercial or financial information.
Exemption 5	Information that concerns communications within or between agencies that are protected by legal privileges, including but are not limited to: 1. attorney-work product privilege 2. attorney-client privilege 3. deliberative process privilege 4. presidential communications privilege.
Exemption 6	Information that, if disclosed, would invade another individual's personal privacy.
Exemption 7	Information compiled for law enforcement purposes if one of the following harms would occur. Law enforcement information is exempt if it: 7(A). could reasonably be expected to interfere with enforcement proceedings 7(B). could deprive a person of a right to a fair trial or an impartial adjudication 7(C). could reasonably be expected to constitute an unwarranted invasion of personal privacy 7(D). could reasonably be expected to disclose the identity of a confidential source 7(E). would disclose techniques and procedures for law enforcement investigations or prosecutions 7(F). could reasonably be expected to endanger the life or physical safety of any individual.
Exemption 8	Information that concerns the supervision of financial institutions.
Exemption 9	Geological information about wells.

[4] Freedom of Information Act of 1966, as amended (United States Code, Title 5, Section 552, Public Information; Agency Rules, Opinions, Orders, Records, and Proceedings, January 16, 2014).

[5] A full history of the amendments is available on the National Security Archive website. See National Security Archive, "FOIA Legislative History," web page, undated.

[6] U.S. Department of Defense Manual 5200.01, Vol. 4, *DoD Information Security Program: Controlled Unclassified Information (CUI)*, February 24, 2012.

among DoD components and between DoD and contractors, consultants, grantees, and other government employees for official DoD business purposes (e.g., have a need to know), the procedure for sharing among approved recipients may not be well known, well defined, or well understood.[7] For example, does one OSD office have the right to access another OSD office's internal personnel rules and practices (exemption 2)? What about direct support contractors with a need to know? Who decides? Given the potential ambiguities and interpretations, it is not surprising that different applications of these exemptions may occur. One issue that was noted in our discussions with OSD is having a single piece of information that in one office has a restriction on sharing but in another office does not have any such restriction. Policy dictates that the decision to share or not share FOUO information is up to the information owner and is based on a demonstrated need to know. One individual may interpret an exemption one way, and another individual interpret a different way.

Other Commonly Used Markings

FOUO is but one marking that is commonly used within OSD. Many other markings are commonly used, such as *government only*, *DoD only*, *pre-decisional*, *source selection sensitive*, and *competition sensitive*, according to the OSD employees we spoke with. However, the origins and application procedures for many of these markings are ambiguous at best and nonexistent at worst. For example, *business sensitive* is commonly applied to information, but we could find no basis for it. We were also unable to identify what type of data has this restriction and why. One plausible hypothesis is that *business sensitive* is a derivative of an FOUO exemption and relates to proprietary information. *Government only* and *DoD only* may be related to distribution statements that are placed on technical documents only, and are referred to as Distribution Statement B, C, D, or E.[8] (For a description of the technical data and policy that govern the sharing of technical data, see Appendix D.) *Pre-decisional* appears to be an exemption that is related to FOIA exemption 5. In all of these cases, the connection between the marking and the basis for the marking (which would describe when and how to apply the marking) is unclear. This lack of clarity exacerbates the problems associated with sharing information.

Given the number of potential markings and the lack of clarity regarding how to implement these markings, it would not be surprising to encounter a piece of information that one believes is improperly marked. If information is improperly marked, getting the markings changed can be challenging. The individual who placed the marking on the information must remove or change the marking.[9] Offices and individuals change over time, sometimes leading to confusion about who has the responsibility and authority for changing the marking if the originator of the document cannot be located or no longer feels responsible. If the originator has left the position or if the office in which the document was originally marked no longer exists, it may be difficult to find others willing to take responsibility for re-marking the docu-

---

[7] Despite the FOUO guidance, many interviewees did not understand when FOUO should be applied and who was able to view FOUO information.

[8] U.S. Department of Defense Instruction (DoDI) 5230.24, *Distribution Statements on Technical Documents*, August 23, 2012.

[9] "The originator of a document is responsible for determining at origination whether the information may qualify for CUI status, and if so, for applying the appropriate CUI markings. . . . (6) The originator or other competent authority (e.g., initial FOIA denial and appellate authorities) shall terminate the FOUO status of specific information when circumstances indicate that the information no longer requires protection from public disclosure" (DoD Manual 5200.01, 2012).

ment. If the individual is still in the office but disagrees with the suggested changes, the process for adjudicating such disagreements is unclear.

However, there are risks even if a document is not marked. In these cases, if there is any concern about the sensitivity of the information, the inclination is to lock down the information, even if there are no markings to restrict sharing. The burden of how to handle the information is then placed on the receiver of the information.

CHAPTER THREE

Practical Issues and Challenges to Sharing Acquisition Data

Some of the challenges to managing classified and CUI information are similar. For both types of information, determinations regarding the appropriate level of protection are required. In some cases the determination is more straightforward than in others. For both CUI and classified information, there are costs and benefits associated with the protection or release of information. In both cases, quantifying the full range of costs is difficult. In the following paragraphs we highlight some of the practical issues and challenges to sharing acquisition data, beginning with those that are common between classified and unclassified information and those that are specific to acquisition information that may fall within CUI.

Classification and Protection Determinations

Determining the level of protection for information is sometimes a challenge. FOUO is considered CUI. However, some interviewees stated that they had challenges determining what is considered FOUO and what is not. In a 2006 report, the Government Accountability Office identified problems with marking of potentially FOUO in DoD:

> [T]he lack of clear policies, effective training, and oversight in DOE's [Department of Energy] and DOD's OUO [Official Use Only] and FOUO programs could result in both over- and underprotection of unclassified yet sensitive government documents. Having clear policies and procedures in place can mitigate the risk of program mismanagement and can help DOE and DOD management assure that OUO or FOUO information is appropriately marked and handled. DOE and DOD have no systemic procedures in place to assure that staff are adequately trained before designating documents OUO or FOUO, nor do they have any means of knowing the extent to which established policies and procedures for making these designations are being complied with.[1]

Costs of Restricting Information

It is difficult to get an understanding of the full cost of restricting, not restricting, or incorrectly restricting information, which can make it difficult to argue for more-open data or more-stringent classification. Undoubtedly, there are direct financial costs. The Information Security Oversight Office (ISOO) reported the "total security classification cost estimate within Gov-

---

[1] U.S. Government Accountability Office, *Testimony Before the Subcommittee on National Security, Emerging Threats, and International Relations, Committee on Government Reform, House of Representatives, Managing Sensitive Information: DOE and DOD Could Improve Their Policies and Oversight*, Washington, D.C., GAO-06-531T, March 14, 2006, p. 8.

ernment for [fiscal year] 2013 is $11.63 billion."[2] Industry is estimated to add an additional $1.07 billion. These costs are for the protection and maintenance of classified information systems, physical security, personnel security (i.e., clearances, access), classification management, declassification, operations security (OPSEC), and training. Another direct cost is the time that people spend attempting to get access to information so they can do their jobs. This can be quite time-consuming in some cases.

Opportunity costs are also incurred by preventing the open flow of information in the functioning and operation of an organization. In fact, excessive classification "prevents federal agencies from sharing information internally[,] . . . making it more difficult to draw connections."[3] Individuals spend time, effort, and money attempting to gain access to classified information. The time and effort spent on attempting to gain access could be applied to more productive endeavors. If access cannot be granted, then inferior information or data may be used to make decisions. While the inefficiency introduced by a lack of information sharing cannot easily be translated to a monetary value, it does have an effect on the operations of an organization. Conversely, the cost of unauthorized disclosure can have significant consequences that are also difficult to quantify.

Problems Accessing Data

To understand how acquisition personnel within OSD are thinking about acquisition data access and management, we conducted a series of semistructured interviews with 67 acquisition professionals to develop a better understanding of the issues with data sharing within OSD, between OSD and the services, and between OSD and support contractors. Presented below are the problems and challenges that were discussed during the interviews. Notably, most interviewees did not mention the statutes or regulations that govern their access and use of data. All were aware of the restrictions associated with specific kinds of data or sensitivity labels.

OSD Functional and Subject-Matter Experts

We discussed with members of each category the types of acquisition data they need, how they are restricted using markings or labels, and the problems the interviewees encounter in getting the data needed.

Types of Data Needed and Classification

Acquisition data needs were extensive but varied based on the mission of the office. Specific areas included cost (e.g., performance, schedule, financial), test (e.g., planning, activities, execution, results), engineering (e.g., schedule, technical and performance parameters, key performance parameters/key system attributes, engineering plans), earned value (e.g., contract data and assessments, supply chain metrics, systems engineering), contract (e.g., competition, termination, funding, small business status, list of contractors), and workforce data (e.g., position

[2] Information Security Oversight Office, *2013 Report to the President*, Washington, D.C.: U.S. National Archives and Records Administration, 2014, p. 22.

[3] Elizabeth Goitein and David M. Shapiro, *Reducing Overclassification Through Accountability*, New York: Brennan Center for Justice, NYU School of Law, 2011, p. 1.

information, qualifications, tenure, assignment, promotion, waivers). The offices also needed to gather other acquisition data to support their analyses.

We also engaged interviewees in a discussion of dissemination constraints and sensitivity labels associated with the data. Nearly all interviewees confirmed that they used proprietary information. Pre-decisional, competition sensitive, business sensitive, source selection sensitive, FOUO, and classified were other markings identified by interviewees.

Problems Accessing Data

When we asked OSD subject-matter experts whether they had problems accessing data to perform their roles and missions, they noted a variety of issues that hampered access, indicating that data are obtainable but not always in a timely or efficient manner. Interviewees noted that data sharing is driven by many factors that are sometimes problematic. Some of the main problems identified included

- latency
- political, structural, and cultural barriers to sharing
- conflicting regulations on proprietary data
- issues with utilizing structured and unstructured information in central repositories
- poor planning.

Latency, or the time gap between when data are generated and when they are made available, was one of the problems identified in our interviews. Interviewees noted that they would like to receive data closer to the time they are generated, giving them a better understanding of what is going on in programs. However, multiple interviewees stated that they understood the reasons behind the latency. Interviewees also noted that it takes longer than they would like to obtain data because the services do not always have enough time to compile and properly characterize the data and prepare them for dissemination, or because the services are cautious about disseminating information.

Another problem cited by interviewees was the inability to get information because of specific chokepoints. Chokepoints could be program offices, prime contractors, or other organizations in the approval chain that are unwilling to share data for any number of reasons. Interviewees in this group mentioned only minimal data-sharing problems between OSD offices. Some noted specific political and structural issues between the services and OSD that inhibited data sharing. For example, leadership may limit dissemination when organizations at lower levels want to share, and vice versa. Structural "stovepipes" in OSD or the services may also inhibit data sharing because acquisition personnel tend to share more readily within their stovepipe or specialty rather than across functional areas. Other barriers to information access are based on personalities and organizational culture. For instance, there may be a request for information, but because the person handling the request (e.g., program office personnel, contractors, service-level management, or other personnel in the data-sharing chain) may not promote sharing or may not understand the urgency of the request, that sharing becomes prohibitively difficult.

Interviewees in the subject-matter expert offices mentioned a variety of problems with using proprietary data. First, there is a lack of clarity in what should be considered proprietary information. Some questioned what is and what is not proprietary when the government is paying for the data. Interviewees mentioned that it is difficult to push back on contractors when something may be improperly labeled PROPIN. Given the restrictions on PROPIN data and the legal liability for the mishandling of such data, some of our interviewees noted that they are more cautious in handling PROPIN materials. Others noted that they understood the restrictions on PROPIN material and were careful about sharing it properly but not overly concerned about legal liability. Another related problem is that NDAs are required of nongovernment personnel handling these data. Interviewees mentioned that it is sometimes difficult to get NDAs signed in a timely fashion (e.g., parties involved want to add additional clauses) or at all if a support contractor is considered to be a competitor of the prime contractor that "owns" the data. They also noted that it is difficult to scale the number of NDAs needed when there are several prime contractors and support contractors who need to work with certain types of data. In cases in which many people are involved in signing NDAs, it can be difficult to keep track of the authorizations.

Interviewees also noted some problems with retrieving information from central repositories. Although they acknowledged that central repositories are beneficial when the data in the repositories are sufficient and in a format suitable for analysis, they also noted several shortcomings:

- There may not be sufficient funding to expand or update the central repositories to fulfill all data needs, which makes the repositories useful for some but not all purposes.
- Access to central repositories may be limited by various business rules.
- Not all personnel who might benefit from them know about central repositories.
- Central repositories tend to limit real-time access to data until the data are properly processed, described, or checked for accuracy. Interviewees said that they sometimes need real-time data.
- There are sometimes duplicative databases, so it is not always clear which structured source is the authoritative one for a specific data set.
- The processes for getting access to central repositories are not always the same (meaning that repositories use a variety of business rules and IT-related measures for access). Some require a Common Access Card (CAC), while others require usernames and logins only. Still others require a CAC and permit access only from a military network. Interviewees noted that it is sometimes easier to go to the program offices than to navigate multiple, somewhat different processes for getting access to repositories.

Finally, interviewees noted that poor planning makes it difficult to retrieve data. Acquisition personnel may not plan ahead or anticipate future data needs. When data-reporting requirements are not properly added to a contract upfront, the government must negotiate with contractors for these data. This can be costly for the government and particularly detrimental when budgets are tight. Furthermore, if contracts contain the wrong clauses, then the government must either modify the contract or find other means to get the data.

OSD OIPT or DAB-Review Offices

Types of Data Needed and Classification

Interviewees in these offices stated that they generally need all kinds of data. Largely, the data pertained to program-planning materials and program documentation. Program offices create and provide most of the data and information that oversight-oriented offices require to perform their mission. The required data are usually maintained at the program office. However, service staffs, OSD-level subject-matter experts (CAPE, PARCA EVM), or the Joint Staff may also maintain them.

Many of our interviewees were part of the OIPT organizations and were therefore heavily involved in DAB meeting preparation and reviews and in developing defense acquisition executive summaries. They also analyze the portfolios of the acquisition programs for which they have oversight responsibility and perform program oversight (by reviewing draft planning and milestone documentation and participating in reviews, technical readiness assessments, and other reviews that reflect on program performance and are part of the oversight process).

Interviewees from the OSD OIPT or DAB-review offices most often mentioned PROPIN, FOUO, and pre-decisional (exemption 5 from FOIA release) dissemination constraints and sensitivity labels. The pre-decisional category was somewhat specific to this group, which has a central role in DAB-related reviews and decision milestones. One interviewee associated the pre-decisional label with a document's timing or stage relative to an upcoming DAB meeting.

Problems Accessing Data

Interviewees from the OSD OIPT and DAB-review offices mentioned encountering a range of data access or handling problems. Lack of access to comptroller databases containing procurement, research, development, testing, and evaluation budget details made it difficult to pull together budgeting information for portfolio analyses. Others noted that programs are more likely to share data and information when things are going well than if a program is encountering execution problems.

Interviewees also noted access problems for support contractors working in the OSD OIPT or DAB-review offices with respect to documents in AIR. Part of the issue is that the programs (or the services more generally), as document owners, can specify access constraints when uploading their documents to AIR, which means that access can be denied by a document owner regardless of the specific rationale for the request. Furthermore, each document or type must be requested individually. The effort needed to access a large number of documents is a disincentive for using the repository. Finally, to make informed decisions about programs, OSD OIPT or DAB-review offices need access to the various data, as required by Department of Defense Interim Instruction 5000.02, before they are archived in AIR.[4] If it takes several months to upload information on a particular program after a major decision, these offices cannot fulfill their missions.

Combining unclassified data that reside on both the Non-Secure Internet Protocol Router Network (NIPRNet) and Secure Internet Protocol Router Network (SIPRNet) can be problematic. If unclassified information is posted on the SIPRNet but is needed on the NIPRNet, there is a complex process for moving the data, leaving analysts to find another, less difficult source. Interviewees also mentioned a lack of data consistency between central repositories; in other words, budget data in DAMIR do not always match comptroller data, making it difficult to compare budget information. Another facet of this same problem is that the comptroller's office does not want to grant access to its internal system, which is another source of budget data. Multiple interviewees mentioned not wanting to provide access to repositories set up for internal office use only. The rationale was that the data reside in multiple formats and they are still raw data. They need to be cleaned—that is, reviewed to ensure that they make sense and

---

[4] U.S. Department of Defense Interim Instruction 5000.02, *Operation of the Defense Acquisition System*, November 25, 2013.

do not contain errors before being shared. CAPE's process for clearing contractors to use its repository (Defense Cost and Resource Center [DCARC]) can be time-consuming, and must be renewed annually or when tasks change. These comments were offered in the context of analyzing portfolios of programs, meaning that data must be pulled on multiple programs. These comments are very similar to those made by personnel from analysis-oriented offices. Studies cannot be started without access to data, often from multiple sources.

Some interviewees discussed the timing of data sharing, noting that the requesting office should wait until the providing office is ready to share. It is easy to see how this might apply to OSD data requests to the services. The chain of command also needs to be respected, which means that unless the requesting office already has an established relationship with the program office, data requests should go through a service liaison office and through the program office, if the data requested are from the prime contractor who is contracted to build the weapon system.

Some interviewees recognized that data must be reviewed for accuracy before sharing to avoid erroneous conclusions by a requester. Prime contractors disclosing information to OSD without first going to the program manager can introduce interpretation problems. Other interviewees acknowledged that OSD offices want to be provided preliminary data and they want to be directly involved with programs all the time, but that is not always appropriate. Sometimes it is more about requests for data that are not routinely collected or generated or for data that may exist but not in the right format to share. Some interviewees from the analysis-oriented offices made similar observations.

Despite the perceived extent of these problems, most interviewees presented them as merely annoying or time-consuming tasks; no OIPT or DAB-review office mentioned any severe access problems that either prevented access to data or hindered the ability to perform required functions to some degree. Most used the data or information available at the time (this response was also similar to observations made by analysts). One interviewee noted that, sometimes, difficulty obtaining data was not really about the data per se but about whether there was an established relationship and trust between the data owner and the requestor.

In general, interviewees in the OSD OIPT or DAB-review offices did not indicate any major differences in data access among the various services. Differences were more likely due to personality or the philosophy of an office's senior leadership with respect to sharing information. Similarly, no differences in data access were noted for different commodity types.

OSD Analysis Offices

Types of Data Needed and Classification

Interviewees in the analysis-oriented offices generally said they need access to the full spectrum of data, cutting across both functional areas and programs, including program documentation, planning materials, briefings, and information that enables an understanding of the cost, schedule, and performance of programs or portfolios of programs. Much of the data and information required to support their analyses are maintained at the program-office level. Many of their analyses require data from multiple program offices, and contacting each custodian would be prohibitively time-consuming and possibly burdensome to the programs. Thus, they rely on central repositories as their primary source of data.

Problems Accessing Data

Problems Generated by Data-Marking Criteria

Handling data marked as PROPIN or FOUO was problematic, according to interviewees in the analysis-oriented offices, in large part because the criteria used to mark documents PROPIN or FOUO were neither transparent nor consistent across types of documents and data owners. The upshot is that different documents with the same information may be marked differently. Several interviewees told stories of presenting unclassified information that had been approved for public release at conferences and later seeing that same information marked as proprietary in a contractor's presentation. Proprietary information is protected by law, but interviewees noted that few understand how the law applies in specific cases. It is not a problem of policy for contractor data. The issue is with the interpretation and application of the Trade Secrets Act.[5] Those who are uncertain about how to apply the PROPIN label may err on the conservative side and not share data. Several interviewees described a general lack of understanding of, or enthusiastic disagreement with, the basis for restricting earned value data as PROPIN or as covered by the Trade Secrets Act. Some argued that these data were not proprietary.

Stovepipes, both within OUSD(AT&L) and among external organizations, were also identified as a constraint to sharing data, including access approval. An organization tends to assist those with the same specific mission or role before assisting those asking for information outside its mission. This seems to be both a structural and cultural issue that interviewees were well aware of, but they believed that it was difficult to change. Related to this issue, interviewees observed that there is no single business chain of command, which creates issues across the different business processes. Interviewees identified four different statutorily defined command chains:

1. OUSD and the Comptroller/Financial Management oversee the execution of funds.
2. OUSD(AT&L) provides oversight and approval of acquisition programs and processes.
3. Only a contracting officer has the authority to sign a contract.
4. Only USD(P&R) can assign billets, which are absolutely necessary to get any work done.

According to interviewees, a lack of collaboration may lead to differences in interpretations of data-access policies and processes.

Some interviewees also noted the lack of standardized markings on documents. Information security requirements can constrain open access. Most interviewees recognized bureaucratic impediments for sharing data, including the fact that only a few people can authorize access, while many can block it. Additionally, program offices generally require some level of approval before they can release data, which delays full disclosure or data sharing outside the office until approvals are completed; it also may result in decisions being made without all desirable information.

Interviewees from the analysis-oriented offices also noted that it is somewhat easier to obtain data from other offices within OSD (despite the stovepipe issue) than it is from the ser-

[5] *Trade Secrets Act* is a catchall term that applies to a series of state and federal laws governing commercial secrets. In contrast, the Uniform Trade Secrets Act is a piece of proposed legislation, subsequently adopted by a majority of states and the District of Columbia, based on language drafted by the Uniform Law Commission in 1985. See Uniform Law Commission, "Legislative Fact Sheet: Trade Secrets Act," web page, 2014.

vices, but this was not a primary concern. Interviewees did not mention differences among the services. However, some interviewees observed that it was more difficult to gain access to data from the space community, within both DoD and industry.

Analysis organizations reported difficulty in getting information from the services when they did not have preexisting contacts. Some interviewees viewed policy, authority, and data sources as fragmented. Organizations external to DoD (FFRDCs and UARCs) can have difficulty obtaining data that require specific kinds of network connections (i.e., access to a .mil network).

Problems Accessing Centralized Data Repositories

Defense acquisition programs generate a significant amount of data or information over their life cycles. Much of this information is useful for execution, oversight, and analysis. Major improvements have been made over the past few decades in the ability of organizations to capture and store program data in electronic format. This information needs to be accessible for future use, and central repositories are the mechanism for capturing acquisition data, storing them, and releasing them for various uses. Here, we describe seven OSD central repositories that can be used to assist in managing various types of acquisition data.[6] This description also provides additional background and details about benefits and challenges to use. The seven repositories are:

- Acquisition Information Repository (AIR)
- Defense Acquisition Management Information Retrieval (DAMIR)
- Defense Automated Cost Information Management System (DACIMS)
- Defense Technical Information Center (DTIC)
- Federal Procurement Data System–Next Generation (FPDS-NG)
- PARCA's Earned Value Management Central Repository (EVM-CR)
- OUSD(AT&L)'s Workforce Data Mart.[7]

The seven repositories contain various types of acquisition data. More specifically, they contain acquisition information from the information requirements defined by DoDI 5000.02.[8] They also include more-detailed cost, budget, earned value, scientific, technical, engineering, contract, and workforce data. The typical procedure is to have a "trusted agent" or government sponsor that will vouch for the need for access to certain information. Government employees always have an easier time getting access than contractors, because government employees are presumed to have a need to know because of their official function and are permitted to access proprietary information. The use of a DoD CAC or public key infrastructure (PKI) is also normally required for access. Users can also get access by having an external certificate authority. Several repositories have classified versions, but we address the unclassified versions only.

[6] We only provide information in this chapter on the unclassified versions of these repositories. For instance, DAMIR has a classified capability, but that is not discussed in this chapter.

[7] Table C.1 in Appendix C provides crossrepository comparisons.

[8] DoDI 5000.2, *Operation of the Defense Acquisition System*, January 7, 2015.

Acquisition Information Repository

AIR is the newest of the repositories in this sample. It was deployed in 2012 by USD(AT&L) and is hosted by DTIC.[9] The purpose of AIR is to support milestone decisionmaking and analysis. The Office of Enterprise Information and OSD Studies developed AIR to consolidate acquisition information required by the current DoDI 5000.02. AIR is a searchable document repository that currently stores more than 300 unclassified Defense Acquisition Executive–level milestone decision documents related to 105 distinct programs.[10]

AIR has one of the stricter access policies. In addition to using a CAC, users must also access the system from government-furnished equipment. This requirement excludes some potential users who are contractors but may not work in a DoD office (e.g., those who work for FFRDCs). The information in AIR covers all of the 46 information requirements in the current DoDI 5000.02. Therefore, there are several different markings, including unclassified, FOUO, pre-decisional, and PROPIN. Given that it is a relatively new repository, content is still being populated, and the repository is still building its user list.

Defense Acquisition Management Information Retrieval

DAMIR was stood up in 2004–2005 to provide enterprise visibility to acquisition program information. It is the authoritative source for current and estimated acquisition category (ACAT) I and IA program costs. DAMIR is well known for being the main repository of selected acquisition reports (SARs) for Congress. It also has begun to add significant capability beyond being a source for document retrieval. Over the past few years, more capability has been added to aid in the analysis of multiple programs across multiple types of acquisition data. It now has more than 6,000 users from OSD, the defense agencies and field activities, FFRDCs, academia, Congress, and the combatant commands.

Defense Automated Cost Information Management System

DACIMS was established in 1998 by OSD CAPE's predecessor, Programs Analysis and Evaluation (PA&E). DACIMS (along with the EVM-CR) is now hosted in its Cost Assessment Data Enterprise using the DCARC portal. DCARC stores various cost data. According to DCARC,

> The primary role of the DCARC is to collect historical and current Major Defense Acquisition Program (MDAP) and Major Automated Information System cost and software resource data in a joint service environment. This data is available for use by authorized government analysts to estimate the cost of ongoing and future government programs.[11]

DACIMS contains around 30,000 contractor cost data reports, software resources data reports, and associated documents.[12] Anyone in DoD can apply for access to DCARC, but some access restrictions may apply. FFRDCs, UARCs, and universities can also apply on an as-needed basis. There are several restrictions on the data in DACIMS, including PROPIN, FOUO, business sensitive, and Not Releasable to Foreign Nationals (NOFORN) (export controlled).

---

[9] Frank Kendall, "Approval of Acquisition Information Repository Policy Memorandum," memorandum, Washington, D.C.: Acquisition, Technology and Logistics, Department of Defense, September 4, 2012.

[10] Kendall, 2012, p. 1.

[11] Defense Cost and Resource Center, homepage, undated(b).

[12] Defense Cost and Resource Center, "About DCARC," web page, undated(a).

Given that DACIMS contains proprietary data, it is one of the repositories that has been trying to come up with a better way to deal with the many NDAs required to access these data.

Defense Technical Information Center

Having started in 1945, the DTIC repository is the oldest of the repositories in this case study. For nearly 70 years, DTIC has grown into a major central repository for DoD scientific and technical information (STI). It has multiple functions, including acquiring, storing, retrieving, and disseminating STI. In addition, DTIC hosts more than 100 DoD websites.[13] Unlike some of the other repositories, DTIC is very large and accessible to anyone through an online portal. In addition to this open site, DTIC has a controlled-access repository that requires a CAC or username and password for access. Authorized DoD and U.S. military employees, U.S. government employees, and government contractors and subcontractors can access the closed area.

Federal Procurement Data System–Next Generation

FPDS-NG is a large repository of federal contract information operated by the General Services Administration (GSA). As is the case with DTIC, a large portion of the repository is searchable by the general public. FPDS-NG "is also relied upon to create recurring and special reports to the President, Congress, Government Accountability Office, federal executive agencies and the general public."[14] DPAP is the DoD representative for the FPDS-NG. It has been in existence (previously, as FPDS) since October 1, 1978, and currently has more than 60,000 users. Users can extract a wide variety of contract information from the site.

Earned Value Management Central Repository

EVM-CR is the authoritative source for earned value data in DoD. EVM-CR is a joint effort between USD(AT&L) and CAPE, and it is managed by AT&L/PARCA. According to its website, the repository provides the centralized reporting, collection, and distribution for key acquisition EVM data.

According to PARCA,

> In July 2007 OSD AT&L released a memo announcing the full implementation across the Department of the Earned Value Management Central Repository (EVM-CR). The EVM-CR provides secure warehousing of Earned Value (EV) data as supplied by the performing contractors on primarily ACAT I Level Programs. Since 2007, various enhancements to the functionality of the EVM-CR have been introduced.[15]

The repository's user access is based on a CAC or trusted certificate. The repository owner also does not allow the different services to see other service data (e.g., the Navy cannot see Army data). The same is true for contractors. This repository is specifically for earned value, so it has a somewhat smaller user base than some of the broader repositories.

---

[13] Shari Pitts, "DTIC Overview," briefing, Defense Technical Information Center, undated, p. 3.

[14] Office of Defense Procurement and Acquisition Policy, "Federal Procurement Data System–Next Generation (FPDS-NG)," web page, last updated February 27, 2014.

[15] Office of the Assistant Secretary of Defense for Acquisition, Performance Assessments and Root Cause Analysis Directorate, *Earned Value Management Central Repository (EVM-CR) Data Quality Dashboard User Guide*, undated, p. 2.

Human Capital Initiative Workforce Data Mart

The Workforce Data Mart has a completely separate data set from the other repositories. The focus is entirely on the acquisition workforce. DAU hosts the repository, but OUSD(AT&L)/HCI directs it. The purpose of the site, according to its guide, is as follows:

> The AT&L DAW [Defense Acquisition Workforce] Data Mart essentially serves many purposes. It de-conflicts workforce members who may be claimed by multiple components to provide the most accurate AT&L [defense acquisition workforce head] count. The business intelligence tools within the Data Mart provide stakeholders such as the military Services, 4th Estate career managers, [functional integrated process teams], and DoD human capital planners with the capability to run reports to conduct analysis and make strategic decisions regarding the workforce. The data in the Data Mart is populated with data defined in the DoDI 5000.55 plus some additional demographic data elements sourced by the Defense Manpower Data Center.[16]

Access is restricted to government workforce managers, functional managers, and non-government employees whose access is sponsored by DoD, because the repository contains sensitive personnel information.

These central repositories offer a number of benefits. One that was identified by multiple interviewees was that they reduced the burden of fulfilling data requests by the program offices. This is an important benefit; interviewees agreed that program offices, which generate significant amounts of acquisition data, cannot fulfill all requests for data and execute acquisition programs properly. A second benefit interviewees cited is that central repositories allow analysts to pull information on a variety of topics and programs from one source. This is particularly helpful for analysts who cover multiple programs or multiple types of acquisition data (e.g., cost, schedule, performance, earned value). In most but not all cases, repositories are set up to provide immediate access, although in practice access requires obtaining appropriate permissions and therefore does take time. Finally, interviewees observed that central repositories provide a unique historical trail of data that may no longer be accessible because of changes in organizational structure (e.g., program offices that no longer exist).

As valuable as the repositories are, the interviewees still noted problems with using them. First, the various repositories have many scanned documents. Depending on the format, scanned documents are difficult to search (i.e., some are images only that have not been converted to searchable text). Second, interviewees noted that many central repositories lack OSD-level pre-MDAP information and testing data. Third, interviewees also agreed that it takes a long time to master using the various central repositories, because the software and structure often differ widely across databases. Because of this, some interviewees reported that they did not access these repositories regularly, or they relied heavily on staff members with better knowledge of the databases. Another concern of interviewees was that there was not a centralized or authoritative process for scrubbing and validating all data in a given repository, which may lead to inconsistencies across repositories.

There are also issues for those who manage the repositories. One is that contractors are not always willing to release the information, so it does not appear in the databases. The

---

[16] Office of the Under Secretary of Defense for Acquisition, Technology and Logistics; Human Capital Initiatives Directorate; and Workforce Management Group, Team B, *Defense Acquisition Workforce Data and Information Consumer Guide*, Version 1.1, April 11, 2013.

owners of those repositories face a myriad of challenges related to sharing, including integrating information assurance and security policies and procedures, along with business rules, into the architecture of the systems. They also must integrate verification of who can and cannot access which data in the systems. Approving access is not a trivial task, with the thousands of potentials users who want access. Another problem identified during our interviews was that the process of retrofitting systems after the introduction of new security policies or business roles tends to be very cumbersome and time-consuming.

Summary

Our interviews indicate that current access to data and information appears inefficient at best and a partial impediment to analysis and oversight at worst. While many government personnel supporting the acquisition process claim to get some of what they need to perform their jobs, they often do not get their first choice of data, and what they do get may not be delivered in a timely fashion. Their task may also be defined based on the information available rather than those data known to not be available. OSD analytic groups and support contractors have unique data-access challenges. Direct support contractors have problems similar to OSD analysts, but these problems can be compounded by regulations and policy that restrict access to certain types of information (especially proprietary data).

CHAPTER FOUR

Proprietary Data: A Case Study

DoD handles different types of data on a regular basis. In this chapter, we look more closely at one category of data that seemed to be of keen interest to acquisition personnel in DoD: proprietary information. Our purpose is to clarify what we mean by *proprietary data*, identify key legal and regulatory regimes that govern the use and protection of proprietary data, highlight recent changes in the regulations concerning the handling of proprietary data, and review a number of notional situations in which the use of proprietary data could cause logistical difficulties for offices whose analysis relies on contractor-provided proprietary information. We include this more detailed look at proprietary data as just one example of the complicated environment that arises when data, regulations, workforce demographics, and the demands of business and policy interact. Our review suggests that further attention ought to be paid to the issues of data availability and efficient access in service of DoD's needs.

What Is Proprietary Information?

PROPIN is a categorization of data based on ownership. Proprietary data are owned by an entity outside DoD. Such entities may be private companies, universities, FFRDCs, or non-profit organizations that carry out their own research and development activities. Furthermore, proprietary data are data that an entity does not wish to disclose freely to the public. Reasons for this preference can range from the data's commercial value to a belief that disclosure would be harmful to current or future business. In short, proprietary data are data that someone has both generated on his or her own and wants to keep private for business reasons. A raft of information can be considered PROPIN: copyrights, patents,[1] trademarks and trade secrets, and business practices, processes, and finances. This information provides a business with a competitive advantage or otherwise contributes to profitability, viability, and success.

DoD has established a definition of *PROPIN* for use by the defense agencies and components. DoDI 5230.24 defines *proprietary information* as follows:

> Information relating to or associated with a company's products, business, or activities, including, but not limited to, financial information; data or statements; trade secrets; product research and development; existing and future product designs and performance specifications; marketing plans or techniques; schematics; client lists; computer programs; pro-

[1] Patents are publicly accessible, but the technology cannot be used without agreement from the patent holder.

cesses; and knowledge that have been *clearly identified and properly marked by the company as "proprietary information," trade secrets, or company confidential information.*[2]

Notable in this definition is the role that the company plays in identifying information as proprietary. The company itself, not DoD, labels information as proprietary and thereby triggers the protections (both DoD and legal protections) that govern its use and distribution beyond DoD.

One further caveat to the definition is worth noting: "The information must have been developed by the company and not be available to the Government or to the public without restriction from another source."[3] Here, the government stresses that the company must have developed the information that it seeks to label as PROPIN. Data or information developed by a third party and subsequently obtained by the company cannot then be passed along to the government with a PROPIN label created wholly by the company. (However, if the third party had originally labeled the information PROPIN, that label might carry forward in some instances.) In short, PROPIN is a self-determined label put on data or other sorts of protectable information by a company to prevent the disclosure of the information to other parties or the public at large.

The labeling of data as PROPIN triggers safeguards and potential legal penalties for the mishandling of the information. When a company has labeled information as PROPIN, the recipient is bound to protect the data from intentional release to the public or other unauthorized users. The reasons for these protections are numerous, but the core sources are found in statutes and regulations. Specifically, proprietary information is governed by the Trade Secrets Act. Something of a misnomer, the Trade Secrets Act is a catchall term that applies to a series of state and federal laws governing commercial secrets. Because DoD deals with companies and nonprofits that may fall under the jurisdiction of state law, both federal and state protections for trade secrets and proprietary information are relevant. The Trade Secrets Act provides for civil penalties for the unlawful disclosure and misappropriation of trade secrets, including injunctive relief through the courts, monetary damages, and the potential award of attorney's fees. Beyond the Trade Secrets Act, the Economic Espionage Act of 1996 further enforces the protection of trade secrets by making their misappropriation a federal offense.[4] Additional legal rules and criminal penalties associated specifically with federal government employees who disclose confidential (nonclassified) information are enumerated in 18 United States Code (U.S.C.) 1905.[5]

Besides these legal restrictions on the release of proprietary data, DoD has also issued rules for the marking and handling of information marked PROPIN. DoD Manual 5200.01, Volume 4, released by USD(I), makes clear that PROPIN is a category of data that carries distributional restrictions. A *distributional restriction* refers to whether the information may be shared with another entity. The PROPIN marking must be placed on every file or document that contains the PROPIN information. DoDI 5230.24, issued by OUSD(AT&L), prescribes a set of distribution statements to be listed on documents within DoD. Distribution Statement B

---

[2] DoDI 5230.24, 2012, p. 29; emphasis added.

[3] DoDI 5230.24, 2012, p. 29.

[4] United States Code, Title 18, Section 1832, Theft of Trade Secrets, January 3, 2012.

[5] United States Code, Title 18, Section 1905, Disclosure of Confidential Information Generally, January 16, 2014.

of the scheduled distribution markings addresses the language to be placed on documents that contain PROPIN data. The language of Distribution Statement B reads, "Distribution authorized to US Government agencies only (PROPIN)(DATE). Other requests for this document shall be referred to (insert controlling DoD Office)."[6]

The language of Distribution Statement B and the required marking of documents containing PROPIN data demonstrate how DoD has internalized the legal restrictions and constraints on sharing these data. If carried out, the restrictions and directions summarized in Distribution Statement B would limit access to government personnel only. As we discuss later in this chapter, however, this limitation to government employees has, in the past, been an obstacle to the timely access to information.

Does the Protection of Proprietary Information Hamstring the Efficient Flow of Data Within DoD?

Despite some progress, the regulation of proprietary information and its handling by government actors remains a stumbling block for some government offices. For offices that are tasked with analysis of contractor-provided data, PROPIN causes particular difficulties when contractors are involved, because they are employees of a commercial firm and may have a conflict of interest if they have access to another firm's proprietary data. While some of the difficulties associated with sharing and protecting technical proprietary data were remedied with a 2013 revision to the Defense Federal Acquisition Regulation Supplement (DFARS) (discussed later in this chapter), many challenges remain. To this end, it is important to review the causes of the difficulties and examine where further regulatory action might be needed to address problems that still remain.

To understand why contractor support caused problems for DoD when handling PROPIN data, we present a series of graphics that depict the flow of information in several scenarios in which the government requires contractor-supplied data. These scenarios are meant to be generic to highlight the broad areas of concern about whether data are available to the government or whether the nature of the government's workforce (i.e., government employee compared with contractor support) in any way hinders timely access to data. These scenarios are notional but reflect the general thrust of the concerns we heard in our discussions with personnel from various offices in DoD.

In Figure 4.1, we depict the flow of data within DoD. DoD regularly and routinely handles PROPIN data in a variety of circumstances. Typically, the government has contracted with an outside provider, which we label as the *prime contractor*, which furnishes the data to the government on a restricted basis. The program office receives these data and works with OSD staff to analyze and examine program progress. The data are furnished by the contractor and, in accordance with DoD policy, must be marked by the contractor as PROPIN. Furthermore, the data must be handled by government employees in both OSD and the program office as stipulated in Distribution Statement B.

In a situation in which only government employees staff the program office and the OSD offices, there is no concern about the handling of PROPIN data so long as the data are not released outside those offices. In these situations, the government can use the data for govern-

[6] DoDI 5230.24, 2012.

Figure 4.1
Problematic Proprietary Data Flows (Scenario 1)

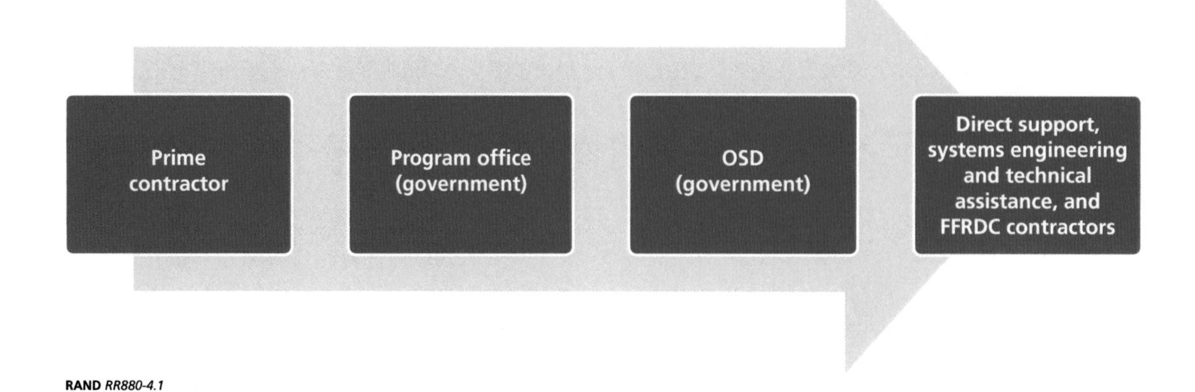

RAND *RR880-4.1*

ment purposes and internal analysis. Recall that Distribution Statement B restricts access to government employees only.

Complications begin to arise, however, when we relax the assumption of government employees constituting the staff of the program and OSD offices. Because the offices rely, sometimes heavily, on the support of contractors to perform analysis, some offices likely seek to distribute analytic tasks calling for use of prime contractor–provided PROPIN data to support contractors. On the right side of Figure 4.1, we identify direct support, systems engineering and technical assistance, and FFRDC contractors to represent the types of secondary contractors that might be called on to carry out these analytic tasks. Because PROPIN rules (prior to 2013, as summarized in Distribution Statement B) stated that the government may not release such information to the public or unauthorized third parties, it would seem that the government may not employ support contractors for analytic tasks using PROPIN data. To the extent that support contractors are not required to use PROPIN data to perform their analytic tasks, there is no problem in the system of data access. However, if offices need—for staffing reasons—to use support contractors for PROPIN-driven research, the prohibition on third-party access presents a difficulty.

In short, when PROPIN data flow from prime contractor through government staff to support contractors, the regulations on the use of the PROPIN label can stifle the flow of data. To use the analytic support contractors fully, the government would need specific permission from the prime contractor to grant access to contractor support staff. Likely, this would take the form of an NDA, a document that would grant access to a specific contractor (or perhaps even a specific employee of a contractor) for specific pieces of information, for a specific period of time. This process, according to our interviews, was a burdensome task and one of questionable efficacy. Of course, prime contractors may perceive support contractors as competitors; under such conditions, prime contractors may deny the support contractors access and thereby restrict data flows.

In a slight variation of the scenario in Figure 4.1, the flow of information may be similarly stymied if access to analytic data stored in a central repository is hampered for PROPIN reasons. In Figure 4.2, we depict the flow of information from prime contractor through government staff to support contractors (who presumably have NDAs that allow the uploading of

Figure 4.2
Problematic Proprietary Data Flows (Scenario 2)

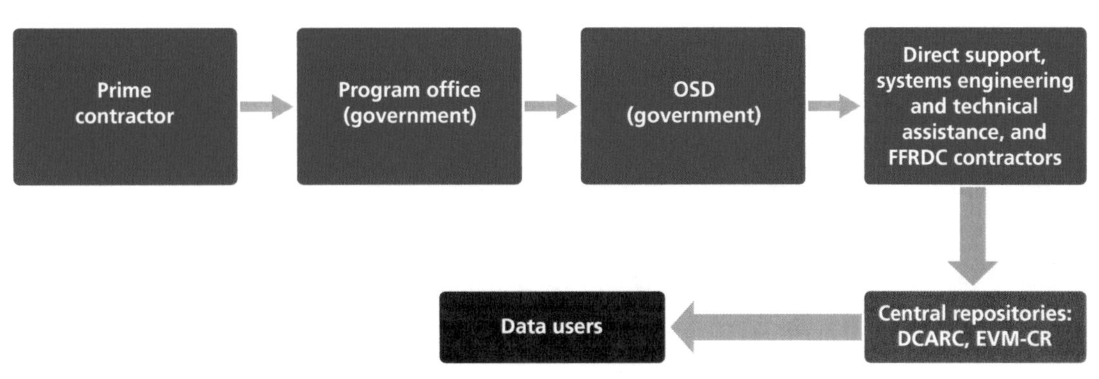

RAND *RR880-4.2*

PROPIN data into a central repository); the data are then compiled into the central repository. Once the PROPIN data are in the central repository, however, questions arise about who—beyond government employees—may have access to the central repository or manage the data in the repository, as well as the way in which access may be granted. Since the central repository may have data from many prime contractors, support contractors seeking access may face significant challenges. If, for example, access to the repository would allow access to all its data, would these contractors need NDAs with every prime contractor whose PROPIN data were stored in the repository (even if the contractor already had one for the specific data needed for the analysis)? Such questions of access, and subsequent questions of monitoring, are not merely academic. Central repositories containing proprietary data abound in DoD (e.g., PARCA's EVM-CR, CAPE's DACIMS). Each of these offices and repositories collects data from multiple prime contractors, processes and prepares the data for storage in the repositories, and performs analysis—potentially involving multiple nongovernment parties.

A somewhat less likely but still possible situation concerning PROPIN data flows can arise if the government itself is kept out of the loop (see Figure 4.3). Consider a situation in which private contractors, both prime and support, solve the data-access problem by coordinating outside the government. If data are transferred from the prime contractor to a support contractor without government involvement, the government's oversight, direction, and visibility into the data may be lost. In this scenario, the government may be able to obtain the analysis it needs in a timely fashion, but it potentially loses its own access to the data. While the concerns of the prime contractor may be mitigated (since the relationship between the support contractor and the PROPIN data is directly managed by the prime contractor), the government's access to the data may be sacrificed. In short, when PROPIN data flow from the prime contractor directly to support contractors (presumably using NDAs between prime contractor and nongovernment staff), government personnel may be excluded.

Faced with concern about government offices being hamstrung by a myriad of NDAs, Congress required a revision to the regulations surrounding the role of government support contractors and access to PROPIN data. As part of the National Defense Authorization Act of 2010, Congress defined government support contractors as those on a contract with the government, "the primary purpose of which was to furnish independent or impartial advice

Figure 4.3
Problematic Proprietary Data Flows (Scenario 3)

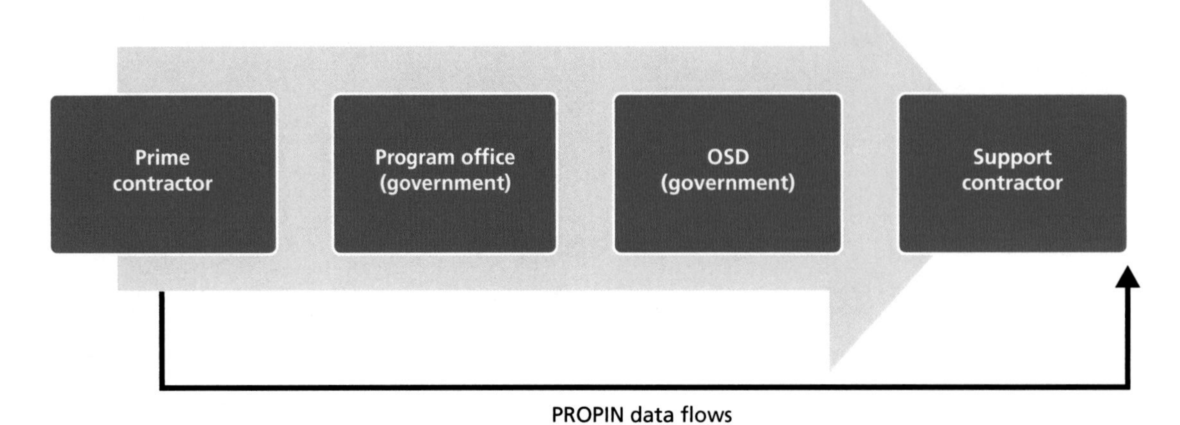

RAND *RR880-4.3*

or technical assistance directly to the government."[7] In our scenarios, the support contractors would likely qualify as government support contractors under the National Defense Authorization Act definition. These newly defined contractors, according to the 2010 law, would have access to proprietary information, subject to legal restrictions regarding nontransmission to the public or unauthorized third parties. This, essentially, would put government support contractors on par with government employees for the purpose of accessing the proprietary information necessary to fulfill the government support contract.

Implementation of the 2010 requirements came in the form of a 2013 revision to the DFARS, which responded to Section 821 of the 2010 National Defense Authorization Act by excepting "government support contractors" from the general prohibition on access to PROPIN data. As a result, government support contractors can now have "access to and use of any technical data delivered under a contract for the sole purpose of furnishing independent and impartial advice or technical assistance directly to the Government in support of the Government's management and oversight of the program or effort to which such technical data relates."[8]

At first blush, it would seem that the revision to the DFARS concerning government support contractors would resolve concerns about data flow and access. Yet the revision pertains to government support contractors and *technical data*. A term defined in law, *technical data* refers to a raft of regulations in DFARS 252.227. We do not delve into the technicalities of the definition here; for our purposes, it will suffice to mention that EVM data—a set of financial data used to measure whether a program's cost and performance are on schedule—fall into a gray area that does not fit squarely within the DFARS definition of *technical data*.

This omission raises precisely the sorts of questions outlined in Figures 4.1 and 4.2, in which support contractors deployed to work with EVM data are prevented from doing so

---

[7] U.S. Congress, 111th Cong., National Defense Authorization Act for Fiscal Year 2010, Washington, D.C., H.R. 2647, Public Law 111–84, October 28, 2009, Section 821.

[8] U.S. Department of Defense, Defense Federal Acquisition Regulation Supplement, Government Support Contractor Access to Technical Data (DFARS 2009-D031), *Federal Register*, Final Rule, May 22, 2013.

because the new regulation does not seem to apply to EVM data (because the law refers only to access to "technical data"). Furthermore, because the nature of EVM data is not clearly established in the regulation, it is not clear whether these are "technical data" or "financial data." Hence, whereas the Federal Acquisition Regulations and DFARS regulations provide clear guidance regarding data rights for technical data—including remedies for inappropriately restricted data and access to technical data by government support contractors—the regulations do not provide corresponding guidance regarding data rights for financial or management information, such as EVM data.

In sum, the PROPIN environment has created a situation whereby the government has initially restricted contractor access to PROPIN data, then subsequently begun a patchwork process of granting access in limited circumstances. But the patchwork process is incomplete. EVM data represent only one of potentially many types of nontechnical data that government offices use. To the extent that these offices rely on contractor support for their data management and analysis, they may be barred from doing so until similar revisions to the Federal Acquisition Regulations and the DFARS are approved.

CHAPTER FIVE

Conclusions and Options

Our findings show that access to data and information is inefficient at best. Many government personnel supporting the acquisition process often do not get their first choice of data, and what they do get may not be delivered in a timely fashion. When they do not get the requested data in a timely fashion or when they do not get the requested data at all, they try to find alternative data and information to perform their assigned duties. However, those data may not be from authoritative sources and may require multiple caveats explaining their limitations. While the consequences of these limitations are undocumented and difficult to assess and quantify, the results of these analyses can be inferior, incomplete, or misleading.

Two groups of people have particular challenges when it comes to gaining access to data required to perform their responsibilities in the support of the Defense Acquisition System: OSD analytic groups and support contractors. These groups have a large variety of data requirements covering many topics in defense acquisition that cross OSD offices and the services. These analysts tend to rely more on a few central repositories as primary sources of information. These repositories may not be able to meet all data needs and do not support exploratory analyses well. OSD analytic groups often do not have access to the originators of the data, which precludes them from going to the primary source. They also tend to have poor visibility of all viable data sources, which encourages inefficient data-seeking practices.

Direct support contractors have problems similar to OSD analysts, but these problems can be compounded by regulations and policies that restrict access to certain types of information (especially nontechnical proprietary data), which introduces extreme inefficiencies. Support contractors require special permissions to view nontechnical proprietary data, such as EVM data or cost performance data. Obtaining these permissions can be a lengthy and cumbersome process. For example, DoD may request that an FFRDC conduct a study on cost growth in MDAPs. The study team would require access to the cost data of every major defense acquisition program. To get this data, the FFRDC would have to establish an NDA with every prime contractor of an MDAP. The prime contractor may require an NDA between the specific individuals requiring access to the data, or may allow for an NDA between the two organizations. In addition, the prime contractor may require an NDA between a specific part of the company and the requesting activity. This process could take months, and may have to be performed for every study requiring this information. This is particularly problematic for organizations that rely heavily on a nongovernment staffing model or that need answers to a particular policy problem quickly. Unfortunately, the DoD staffing model falsely assumes efficient data access.

Access issues and challenges occur for a variety of reasons.

- Data-access policy is highly decentralized, not well known, and subject to interpretation.
- The markings applied, or not applied, to unclassified information play a significant role in access. It is up to the owner or creator of a document to determine what protections and markings are required. However, marking criteria are not always clear or consistently applied. In fact, management and handling procedures for many commonly used markings are not clearly described anywhere. If information is improperly marked, getting the labels changed can be difficult. Offices and individuals change over time, leaving it unclear who the originator was. If the originator has left the position or an office has closed, then others may not want to take on the responsibility of re-marking the document. When information is not marked, the burden of handling decisions is placed on the receiver of the information.
- Theoretically and practically, there are institutional and cultural barriers to sharing. For example, interviewees pointed to the stovepiped structure of DoD and how it limits visibility and the sharing of data and information. Institutional structure and bureaucratic incentives to restrict data access are exacerbated by policy and guidance to protect information. The result is a strong conservative bias in labeling and a reluctance to share. Some interviewees and the literature on change management describe how a lack of trust and established relationships can hinder sharing. The literature and some interviewees also identified how important senior-leadership support is to the success of any corporate initiative, including data-sharing initiatives.

Options to Address Proprietary Data-Access Problem

OUSD(AT&L) has options that it may pursue to mitigate some of these issues, but each option requires further analysis and staff coordination. To resolve the problem of access to proprietary data, the following steps could be taken:

- OUSD(AT&L) could seek additional billets and insource any functions that require access to proprietary data, thereby circumventing the burdensome nondisclosure process that is currently in place. However, this would require Office of Personnel Management (OPM) and congressional support, which can take time and effort to develop.
- OUSD(AT&L) could seek relief through a reallocation of billets to functions that currently require access to proprietary information. This would require cross-organizational prioritization in a structure where no office wants to give up its limited billets.
- Establishing general access to nontechnical data for all direct support contractors is another option, but it requires legislative or contractual changes. For example, there is currently legislation, 10 U.S.C. 129d, that allows disclosure of sensitive information to litigation support contractors.[1] Similar legislation might be pursued for all support contractors. Alternatively, or in conjunction with legislative change, relief could be sought through the addition of contractual language that would be placed on all DoD acquisition contracts that grant support contractors restricted access to the resulting data. The

---

[1] United States Code, Title 10, Section 129d, Disclosure to Litigation Support Contractors, January 16, 2014.

Table 5.1
Options to Address Proprietary Data-Access Problem and Implications

Options	Implications
Insource further	• OPM and Congressional engagement
Reallocate billets to tasks involving PROPIN	• Cross-organizational prioritization • Pervasive • Billets limited
Establish general access for contractors	• Legislative or contractual options
Establish less conservative ruling on NDAs	• Still requires data-access controls by company

direct support contractors who receive the data would have to demonstrate company firewalls, training, personal agreements, and need to know akin to those for classified information.

- Finally, the government could seek an alternative ruling on the nondisclosure requirements, whereby blanket NDAs could be signed between the government and a direct support organization, or a company and a direct support organization, to cover multiple tasks. Each option would require further analysis and coordination with the Department of Defense Office of the General Counsel and DPAP (and Congress in the first option).

Table 5.1 summarizes the potential options.

Options to Address Policy Confusion

There are also several options to address the confusion regarding policies on sharing acquisition data. AT&L could create and maintain a central, authoritative online resource that references all relevant guidance on information management, handling, access, and release for acquisition data. This would provide a central location for individuals to seek assistance. AT&L could also consider additional training for its staff on the identification and protection of data. This could be an annual online training for all AT&L staff and contractors.

Additional guidance on how to determine whether information is FOUO or proprietary could be provided to AT&L staff. The guidance should provide specific examples of information that is considered protected by these markings, guidelines for determining whether or not specific information qualifies, and details regarding handling procedures for this information—including access privileges.

Directives and incentives could be established so that markings that appear to be incorrect are challenged and not taken only on a company or individual's claim. A process and guidance to assess the validity of a marking, to include proprietary data, could be developed and implemented.

There are important reasons for restricting access that require balancing control with granting more access. In information assurance and security policy, there is an understanding that no individual should have unfettered access to all data. Given the inherent complexity in securing data and sharing data, any solutions to problems associated with data-sharing difficulties should be well thought out to avoid the multitude of unintended consequences that could arise.

APPENDIX A

DoD Policies Affecting Data Sharing

One of the priorities of this study was to identify and reference major policies that affect data sharing. Our search resulted in the list of policies, presented in Table A.1, that affect acquisition data sharing. Given the vast array of policies, it is possible that additional aspects of data sharing may be embedded in other policies, but these appear to be the major ones. We included both the policies that we based our analysis on and the names of the revised policies since we conducted the analysis.

Table A.1
DoD Policies Affecting Acquisition Data Sharing

Policy Authority	Policy Number	Subject	Date
President of the United States (POTUS)	Executive Order (EO) 13556	*Controlled Unclassified Information*	11/4/2010
	EO 13526	*Classified National Security Information*	12/29/2009
USD(AT&L)	Interim DoDI 5000.02	*Operation of the Defense Acquisition System*	11/25/2013
	Revised: DoDI 5000.02	*Operation of the Defense Acquisition System*	Revised: 1/7/2015
	DoDI 5230.24	*Distribution Statements on Technical Documents*	8/23/2012
	Department of Defense Directive (DoDD) 5000.01	*The Defense Acquisition System*	5/12/2003
	DoDI 5000.55	*Reporting Management Information on DoD Military and Civilian Acquisition Personnel and Positions*	11/1/1991
	DoDI 5230.27	*Presentation of DoD-Related Scientific and Technical Papers at Meetings*	10/6/1987
	DoDI 5535.11	*Availability of Samples, Drawings, Information, Equipment, Materials, and Certain Services to Non-DoD Persons and Entities*	3/19/2012
	DoDI 3200.12	*DoD Scientific and Technical Information Program (STIP)*	8/22/2013
	DoDI 2015.4	*Defense Research, Development, Test and Evaluation Information Exchange Program (IEP)*	2/7/2002

Table A.1—Continued

Policy Authority	Policy Number	Subject	Date
USD(AT&L)	DoDD 5230.25	*Withholding of Unclassified Technical Data from Public Disclosure*	8/18/1995
	DoDI 3200.14	*Principles and Operational Parameters of the DoD Scientific and Technical Information Program*	6/28/2001
Chairman of the Joint Chiefs of Staff	Chairman of the Joint Chiefs of Staff Instruction 6510.01F	*Information Assurance (IA) and Support to Computer Network Defense (CND)*	2/9/2011 Revised: 10/10/2013
USD(I)	DoD Manual 5200.01, Vol. 4	*DoD Information Security Program: Controlled Unclassified Information (CUI)*	2/24/2012
	DoD Manual 5200.01, Vol. 2	*DoD Information Security Program: Marking of Classified Information*	3/19/2013
	DoD Manual 5200.01, Vol. 3	*DoD Information Security Program: Protection of Classified Information*	3/19/2013
	DoDI 5200.01	*DoD Information Security Program and Protection of Sensitive Compartmented Information*	10/9/2008 Revised: 6/13/2011
	DoDI 5200.39	*Critical Program Information (CPI) Protection Within the Department of Defense*	12/28/2010
	DoDI 5205.08	*Access to Classified Cryptographic Information*	11/8/2007
	DoDI 5210.02	*Access to and Dissemination of Restricted Data and Formerly Restricted Data*	6/3/2011
	DoD Manual 5200.01, Vol. 1	*DoD Information Security Program: Overview, Classification, and Declassification*	2/24/2012
USD(P&R)	DoDI 1336.08	*Military Human Resource Records Life Cycle Management*	11/13/2009
	DoDI 1312.01	*Department of Defense Occupational Information Collection and Reporting*	1/28/2013
USD(P)	DoDD 5200.27	*Acquisition of Information Concerning Persons and Organizations Not Affiliated with the Department of Defense*	1/7/1980
	DoDD 5230.11	*Disclosure of Classified Military Information to Foreign Governments and International Organizations*	6/16/1992
	DoDI 2040.02	*International Transfers of Technology, Articles, and Services*	7/10/2008 Revised: 3/27/2014
	DoDI 3025.19	*Procedures for Sharing Information with and Providing Support to the U.S. Secret Service (USSS), Department of Homeland Security (DHS)*	11/29/2011

Table A.1—Continued

Policy Authority	Policy Number	Subject	Date
Office of the Deputy Chief Management Officer (DCMO)	DoDD 5400.11	*DoD Privacy Program*	09/01/2011 Revised: 10/29/2014
	DoDI 5230.29	*Security and Policy Review of DoD Information for Public Release*	1/8/2009 Revised: 8/13/2014
ASD(HA)	DoDI 6040.40	*Military Health System Data Quality Management Control Procedures*	11/26/2002
ASD(LA)	DoDI 5400.04	*Provision of Information to Congress*	3/17/2009
ASD, Logistics and Materiel Readiness	DoD Manual 5010.12-M	*Procedures for the Acquisition and Management of Technical Data*	5/14/1993
DoD CIO	DoDD 5015.2	*DoD Records Management Program*	3/6/2000
	DoD Guide 8320.02-G	*Guidance for Implementing Net-Centric Data Sharing*	4/12/2006
	DoDI 8110.1	*Multinational Information Sharing Networks Implementation*	2/6/2004
	DoDD 8000.01	*Management of the Department of Defense Information Enterprise*	2/10/2009
	DoDD 8500.01E	*Information Assurance (IA)*	4/23/2007
	DoD Manual 8400.01-M	*Procedures for Ensuring the Accessibility of Electronic and Information Technology (E&IT) Procured by DoD Organizations*	6/3/2011
	DoDI 5200.44	*Protection of Mission Critical Functions to Achieve Trusted Systems and Networks (TSN)*	11/5/2012
	DoDI 8510.01	*DoD Information Assurance Certification and Accreditation Process (DIACAP)* Revised: *Risk Management Framework (RMF) for DoD Information Technology (IT)*	11/28/2007 Revised: 3/12/2014
	DoDI 8520.02	*Public Key Infrastructure (PKI) and Public Key (PK) Enabling*	5/24/2011
	DoDI 8320.02	*Sharing Data, Information, and Information Technology (IT) Services in the Department of Defense*	8/5/2013
	DoDI 8520.03	*Identity Authentication for Information Systems*	5/13/2011
	DoDI 8550.01	*DoD Internet Services and Internet-Based Capabilities*	9/11/2012
	DoDI 8910.01	*Information Collection and Reporting*	3/6/2007 Revised: 5/19/2014

Table A.1—Continued

Policy Authority	Policy Number	Subject	Date
DoD CIO	DoDD 4630.05	*Interoperability and Supportability of Information Technology (IT) and National Security Systems (NSS)*	4/23/2007
	Revised: DoDI 8330.01	Revised: *Interoperability of Information Technology (IT), Including National Security Systems (NSS)*	Revised: 5/21/2014
	DoDI 8500.2	*Information Assurance (IA) Implementation*	2/6/2003
	Revised: DoDI 8500.01	Revised: *Cybersecurity*	Revised: 3/14/2014
	DoDI 8582.01	*Security of Unclassified DoD Information on Non-DoD Information Systems*	6/6/2012
ASD(PA)	DoDI 5040.02	*Visual Information (VI)*	10/27/2011
Director, CAPE	DoDI 8260.2	*Implementation of Data Collection, Development, and Management for Strategic Analyses*	1/21/2003
	DoD Manual 5000.04-M-1	*Cost and Software Data Reporting (CSDR) Manual*	11/4/2011
Director, A&M	DoDD 5230.09	*Clearance of DoD Information for Public Release*	8/22/2008
	DoD 8910.1-M	*Department of Defense Procedures for Management of Information Requirements*	6/30/1998
	Revised: DoD Manual 8910.01, Vol. 1	Revised: *DoD Information Collections Manual: Procedures for DoD Internal Information Collections*	Revised: 6/30/2014
	DoD Regulation 5400.7R	*DoD Freedom of Information Act Program*	9/1998
	DoD Administrative Instruction 101	*Personnel and Data Management Information Reporting Policies and Procedures for Implementation of the Defense Acquisition Workforce Improvement Act (DAWIA)*	7/20/2012
	DoD Administrative Instruction 15	*OSD Records and Information Management Program*	5/3/2013
	DoD Administrative Instruction 56	*Management of Information Technology (IT) Enterprise Resources and Services for OSD, Washington Headquarters Services (WHS), and Pentagon Force Protection Agency (PFPA)*	4/29/2013

APPENDIX B

Discussions with OSD

To understand how acquisition personnel within OSD view acquisition data access and management, we conducted a series of targeted interviews. We interviewed acquisition personnel from OSD and several from the services. In all, we spoke with 67 acquisition professionals (both government and contractors) to develop a better understanding of the issues with data sharing within OSD, between OSD and the services, and between OSD and support contractors. The interviews focused almost entirely on OUSD(AT&L) because we wanted to start the process of understanding data-sharing issues within OUSD(AT&L) before moving to other perspectives. We also included CAPE, which is not within OUSD(AT&L), because of its significant cost-assessment role in the defense acquisition process and thus heavy use of acquisition data.[1] This appendix shares some observations from these interviewees that fall generally into the following categories:

- data needs
- data flows
- interviewees' recommendations for improving data sharing.

The problems uncovered in the interviews with data sharing were previously presented in Chapter Two, and are illustrative of issues that the offices who agreed to talk to us are experiencing.

We initially contacted a total of 74 personnel in OSD (68) and the services (six), of whom 42 agreed to talk to us—a nearly 60-percent response rate to our request for interviews. Of the 67 personnel ultimately interviewed, some were contacted by the study team and others provided additional information about data sharing at the request of a contacted interviewee. We talked to acquisition personnel from ten OSD offices, the Army, Navy, and Air Force. Table B.1 shows the offices represented in our interviews.

While we did not specifically address data-access and -sharing issues between OSD and the services, we interviewed acquisition executive representatives from the services to capture their perspective. The service cost centers represent service-level organizations with subject-matter expertise that frequently interact with OSD organizations.

[1] Other offices that may play a role in data sharing and management but were not interviewed for this study included the Office of the Under Secretary of Defense Comptroller/Chief Financial Officer, Office of the Under Secretary of Defense for Personnel and Readiness, Office of the Under Secretary of Defense for Policy, Office of the Assistant Secretary of Defense for Networks and Information Integration/DoD CIO, Office of the Assistant Secretary of Defense for Research and Engineering, and Office of the Director of Operational Test and Evaluation. We began with a narrower sample within OUSD(AT&L) to start to understand data-sharing problems, and to maintain a manageable scope. Future analysis in this area would benefit from in-depth discussions with acquisition personnel in the services.

Table B.1
Offices Represented in the Interview Sample

Office of the Secretary of Defense
• Acquisition Resources and Analysis (ARA) Directorate • Performance Assessments and Root Cause Analyses (PARCA) • Defense Technical Information Center (DTIC) • Defense Acquisition University (DAU) • Human Capital Initiatives (HCI) Directorate • Defense Procurement and Acquisition Policy (DPAP) Directorate • Overarching Integrated Project Teams (OIPTs): Tactical Warfare Systems; Space, Strategic, and Intel Systems; and Command, Control, and Communication, Cyber, and Business Systems • Systems Engineering (SE) • Developmental Test and Evaluation (DT) • Office of Cost Assessment and Program Evaluation (CAPE)
Services
• Air Force Acquisition • Air Force Cost Analysis Agency • Navy Acquisition • Naval Center for Cost Analysis • Army Acquisition
FFRDCs
• Institute for Defense Analyses • RAND

We developed an interview protocol in advance so that the interview process would focus on similar topics across interviews. The interview questions presented below served as a guideline for RAND interviewers. Ultimately, the questions were tailored slightly, depending on the role and mission of the interviewee. For example, governance and maintainer offices are different from creators/owners and users.

- What is your office's role in the acquisition process?
- In that capacity, what information/data do you (your office) need?
- Where does that information reside (e.g., in a database, central repository, or document or with an office or person)?
- Who "owns" the data?
- Who maintains the data (repository)?
- What statute, regulation, policy, or guidance governs access and use?
- How is access to the information/data achieved?
- What dissemination constraints and sensitivity labels are on the data or information?
- What access or data-handling problems have been encountered?
- How have those problems been resolved?
- Do you have any recommendations for improving access to data within OSD?
- Are you getting all the data you need to do your job?
- Is there anyone else in your office with whom we should meet?

Interviewee Responses, by Mission and Role

Over the course of this study, we found that interviewees with similar missions and roles had similar data needs. They also obtained data in similar ways. This section describes the data needs of various groups, how the data flow from one organization to another, and the problems interviewees experienced in getting data. The section also presents interviewee recommendations by mission or role.

OSD Functional and Subject-Matter Experts

We conducted 13 interviews with a total of 19 acquisition professionals in the offices of OSD functional and subject-matter experts.

Data Needs, Labels, and Policy

Among this group of offices, data needs were extensive; however, each office was narrowly focused on data directly related to its subject-matter expertise: cost, test, engineering, earned value, contract, or workforce data. The offices also needed to pull in other acquisition data that would help support their analyses. Table B.2 lists these offices' wide range of data needs. This information was provided to us during our interviews and should not be viewed as inclusive of the data needs of these offices.

We also engaged interviewees in a discussion of dissemination constraints and sensitivity labels associated with the data. Nearly all interviewees confirmed that they used proprietary information. Pre-decisional, competition sensitive, business sensitive, source selection sensitive, FOUO, and classified were other markings identified by interviewees.

Interviewees were asked about statutes, regulations, policies, or guidance governing access to and use of data for their jobs. They generally discussed policy in terms of which policies gave them the authority to collect and manage data or which policy may have been creating problems in their efforts to access and manage acquisition data. Interviewees frequently mentioned

Table B.2
Data Needs of OSD Functional and Subject-Matter Experts

Main Area for Data Needs	More-Specific Data Needs
Cost	Technical, performance, schedule, business financial, and other program data
Test	Test activities, test planning, test execution, and test results; prior studies on testing; schedule; acquisition planning; acquisition decisions; any additional data in the test evaluation master plan; and parametric data from similar systems to develop test plans
Engineering	EVM, schedule, technical and performance parameters, key performance parameters/key system attributes, engineering plans, technical schedule (reviews, tests, events), planning, progress to plan, other information in the systems engineering plan, and software code
Earned value	Contract data and assessments, supply chain metrics, and systems engineering
Contract	Competition, termination, funding, small business status, list of contractors, and all other aspects of contracting information
Workforce	Position information, qualifications, tenure, assignment, promotion, and waivers and exceptions

statutory authority as the driving force behind their data collection but said little about the policies that were causing problems. Various portions of Title 10 of the U.S.C., especially 10 U.S.C. 131(d),[2] provide the legal authority for OSD personnel to collect certain types of acquisition data from program offices and prime contractors. Interviewees also mentioned Public Law 111-23, the Weapon Systems Acquisition Reform Act of 2009, which also carries legal authority for OSD offices to perform certain duties. The act created multiple offices within OSD: CAPE, the Office of the Deputy Assistant Secretary of Defense for Developmental Test and Evaluation, the Office of the Deputy Assistant Secretary of Defense for Systems Engineering, and PARCA. Each office has been given various authorities to collect data and inform Congress of the results of the data collection. Interviewees also mentioned 10 U.S.C. 1761 (Management Information System).[3] This statute calls for data to be collected on personnel in acquisition positions in a standardized format. In addition, the Defense Acquisition Workforce Improvement Act (10 U.S.C., Chapter 87) was also cited with regard to the collection of acquisition personnel data. Outside of these policies, interviewees said that they were able to obtain data based on line items in contracts and memoranda of understanding between parties.

Data Flows

The offices of OSD functional and subject-matter experts frequently retrieved data directly from program offices and prime contractors. They also retrieved data from service-level repositories or other entities in the services (e.g., testing agencies). Functional and subject-matter experts had several reasons for not using repositories as their first choice for data. A few of those reasons are that some of the data do not reside in repositories, as in the case of testing. Experts also need advance drafts of documents that are not entered into repositories until after they are finalized. This was generally a secondary mechanism for getting data, except in one or two instances in which OSD offices pulled information from service-level repositories. Interviewees also acknowledged that they sought data from their colleagues in other OSD offices. Likewise, they used the following OSD central repositories as additional sources for obtaining acquisition data: PARCA's EVM repository, OUSD(AT&L)'s Workforce Data Mart, FPDS-NG, DTIC, DAMIR, DCARC, and AIR.

Interviewee Recommendations

In this section, we share some recommendations from our interviewees in the offices of OSD functional and subject-matter experts. It is important to note that these recommendations have not been independently vetted and assessed for benefits, costs, and legal or regulatory hurdles by RAND or DoD. Some of the recommendations may already have been implemented in some parts of DoD but not in others.

Interviewees had some general recommendations regarding data access and sharing. First, they mentioned that acquisition personnel could establish personal relationships with people who should be consulted when looking for data (e.g., maintaining close relationships with program offices or prime contractors). By establishing personal relationships, they can avoid some of the apprehension involved with sharing data with unfamiliar government employees

[2] "The Secretary of each military department, and the civilian employees and members of the armed forces under the jurisdiction of the Secretary, shall cooperate fully with personnel of the Office of the Secretary of Defense to achieve efficient administration of the Department of Defense and to carry out effectively the authority, direction, and control of the Secretary of Defense." United States Code, Title 10, Section 131, Office of the Secretary of Defense, January 7, 2011.

[3] United States Code, Title 10, Section 1761, Management Information System, January 16, 2014.

or contractors. This approach, for example, would ensure that the recipient understands the limitations and characteristics of the data to avoid mischaracterizations. Interviewees also recommended sharing preliminary information with leadership as early as possible rather than waiting for comprehensive information. Additionally, they said that leaders need to understand the data available to them and their shortfalls and that data requirements and quality should be continuously examined.

Within OSD, interviewees mentioned that it is essential for personnel to understand the duties of their OSD colleagues. This would help establish more relationships to turn to to retrieve or share data. In this same respect, interviewees mentioned using decision forums (e.g., OIPT) as a mechanism for sharing or discussing problems. One other popular recommendation was for a "desk guide" that links documents and data to organizations, given that DoD is large and that visibility (or awareness) of all data is difficult.

There were multiple recommendations regarding proprietary data. Interviewees stressed the importance of ensuring that all parties understand NDAs, a requirement for proprietary data to be shared with nongovernment employees. Understanding how NDAs are used to obtain these data may improve the efficiency of the process for obtaining these agreements. Interviewees also needed additional guidance from the Office of the General Counsel on how to handle PROPIN data. The guidance needs to take into account practical considerations for how DoD is structured and managed, beyond a simple interpretation of the law. Finally, it was recommended that there be a change to statutes or the Federal Acquisition Regulations to allow contractor access with an overarching NDA and sponsor in place. For example, this might be accomplished through a clause in the contract granting access to that company's PROPIN for all support contractors. DoD would then train each individual on safeguards and could disclose who has access to what data if asked.

The majority of the interviewees in these offices expressed a need for better planning. Some recommended that everything relating to data rights and data be included in contracts (e.g., a contract line item number or clause), along with NDAs, to facilitate the process of receiving PROPIN. Others mentioned early engagement in the request-for-proposal phase and a requirement in the proposal or contract for the acquisition data to be placed in a central repository within DoD.

OSD OIPT or DAB-Review Offices

We interviewed 12 people in five OSD oversight offices. OIPT or DAB-review offices have strong roles in the formal acquisition process. They help a program navigate the oversight process and provide analysis and advice to senior leaders from their unique (OSD) perspective based on direct interaction with program offices. They touch on the range of functional topics (e.g., management, contracting, sustainment, testing), as well as acquisition oversight. They often have some insight into the requirements and budgeting aspects of the overall acquisition process. Finally, they identify and raise issues that program offices need to address, and they provide advice to the milestone decision authority on a program's readiness for review or the requirements to pass a technical or management milestone.

Data Needs, Labels, and Policy

Some of these offices are data creators or owners in their own right, in addition to users of data from other organizations. Interviewees in the OSD OIPT or DAB-review offices stated that they generally need all kinds of data—largely program planning materials and program documentation. Program offices create and provide most of the data and information that oversight-oriented offices require to perform their mission. The required data are usually maintained at the program office, though they may be maintained by service staffs, OSD-level subject-matter experts (CAPE, PARCA EVM), or the Joint Staff.

Occasionally, program offices allow selected OSD action officers from the OIPT offices direct access to a program's integrated data environment, which the program offices use to create, manage, share, and store program information. Unlike the subject-matter–expert or analysis-oriented offices, the OSD OIPT or DAB-review offices do not generally maintain repositories of data. The exception would be action officers, who are responsible for specific programs. Action officers maintain repositories of program data for their use on their computers or the office's shared drive.

Many of our interviewees were part of the OIPT organizations and were therefore heavily involved in DAB meeting preparation and reviews and in developing defense acquisition executive summaries. Such personnel also analyze the portfolios of acquisition programs for which they have responsibility for oversight and perform program oversight by reviewing draft program planning and milestone documentation and participating in program reviews, technical readiness assessments, and other reviews that reflect on program performance and are part of the oversight process. They ensure that programs are ready for DAB milestone reviews and coordinate the information package for those milestone decisions.

Most interviewees did not mention the statutes or regulations that govern their access and use of data. All were aware of the restrictions associated with specific kinds of data or sensitivity labels, usually based on their own or their colleagues' experiences.

Interviewees from the OSD OIPT or DAB-review offices most often mentioned PROPIN, FOUO, and pre-decisional (exemption 5 from FOIA release) dissemination constraints and sensitivity labels. The pre-decisional category was somewhat specific to this group, which has a central role in DAB-related reviews and decision milestones. One interviewee associated the pre-decisional label with a document's timing or stage relative to an upcoming DAB meeting.

Data Flows

The majority of OSD OIPT or DAB-review interviewees stated that data and information are accessed formally through a program office, service staff, and, occasionally, a central repository. Some also noted the importance of informal approaches to access, particularly personal relationships with program or service personnel built up over years of interaction.

Interviewee Recommendations

Here, we present recommendations provided by interviewees in OSD OIPT or DAB-review roles. As noted earlier, these recommendations are not based on RAND analysis. According to our interviewees, data-access issues are often resolved by establishing and maintaining good personal relationships. Occasionally, issues need to be raised to higher levels for a decision. One interesting solution involved requiring the program office and the requesting office to make a joint presentation to a joint audience of senior decisionmakers to gain access to the information.

Other recommendations for improving data access from an OSD perspective included the following:

- Maintain good communication and build trust over time.

- Work in multidisciplinary teams.
- Develop a short, simple rulebook for all data access and information sharing that is underpinned by the philosophy of enabling.
- Be aware of policy decisions embedded in the technology/structural decisions of data systems. Some decisions about IT or information security have both implicit and explicit effects on the ability to access and share data.
- Discern which data are truly important; do not collect data for data's sake. Senior leaders should decide which data are truly important to inform their decisions and restrain any additional data requests.
- Reduce the volume and amount of data requests by coordinating among requestors. This recommendation addresses the fact that multiple OSD offices often ask program offices for the same information. Similarly, requestors can first ask whether data have been collected and in what format they are available.
- Educate leadership about which data are available and make them available in one place (e.g., a dashboard).
- Grant access to the data owner's shared drive more often.
- Align data access permissions with the requestor's security clearance and need to know rather than allowing permission to be established by data owners.
- Provide additional training to raise the visibility of data sources and the procedures for accessing data. Training might also be offered on the use of the different central repositories, since software tends to be unique to a particular repository.
- One idea that several interviewees expressed in one way or another was the need for a set of business rules to enable the free flow of summary or key information but that also allowed enough distance between the program office and OSD to let each do its job.
- Similarly, interviewees expressed the need to cut down on the cultural incentive to hold on to information until the last moment. This would require addressing the trust issues that constrain a program office's or other data owner's willingness to share information.
- Several interviewees stated that while they believed that information should be more widely shared in general, sharing unprocessed data was usually not appropriate.

OSD Analysis Offices

We interviewed 13 people in four offices involved with crosscutting analyses. The main role of the analysis-oriented offices is to support senior leaders in their decisionmaking and governance role by providing analyses of acquisition issues. Some of these activities might be special projects or tasks assigned by senior leadership, and others are directly related to a primary task of the organization in which these personnel work (e.g., PARCA and root cause analyses). Education and training was also mentioned as an adjunct role by some interviewees.

Data Needs, Labels, and Policy

Interviewees in the analysis-oriented offices generally saw themselves as requiring access to the full spectrum of data, cutting across both functional areas and programs, including program documentation, planning materials, briefings, and information that enables an understanding of the cost, schedule, and performance of programs or portfolios of programs.

Interviewees from analysis-oriented offices recognized that much of the data and information required for their analyses in support of the acquisition process and senior decisionmakers are maintained at the program-office level. However, because many of their analyses require data from multiple program offices, contacting each custodian would be prohibitively time-consuming and possibly burdensome to the programs. These offices also do not have the routine interactions with program offices that subject-matter–expert and OSD OIPT or DAB-review organizations have. They therefore rely on central repositories as their primary source of data to a greater extent than either of the other types of offices.

The data sources or central repositories specifically mentioned include EVM, SARs, defense acquisition executive summaries, other data in DAMIR, workforce training data (from the DAU and OUSD[AT&L] HCI Directorate Workforce Data Mart), cost and software data reporting and other cost and business information in DCARC, and contract information in FPDS-NG. One office mentioned using the comptroller's classified network for budget data. Interviewees mentioned directly contacting program offices as a last resort.

The central repositories mentioned most often are DAMIR, AIR, DCARC, and EVM. FPDS-NG, the comptroller's budget database, and HCI's Workforce Data Mart are more specialized and were mentioned less often by interviewees from analysis-oriented offices. ARA, CAPE, DPAP, PARCA, the comptroller, and HCI maintain the central repositories.

The data are used largely to assess the performance of the acquisition system, in whole or in part. They are used in analyses that range from supporting decisionmaking (e.g., DAB milestones) to broad characterizations (portfolio analysis or trends over time), to responding to very specific questions from senior leadership or Congress. Some analyses are also posted for more-general use by acquisition professionals.

As was the case with the other two groups of OSD offices, most interviewees in the analysis-oriented offices did not mention specific statutes, regulations, policies, or guidance governing access to and use of data. However, all were aware that such statutes and regulations existed and that different types of data (or information), or data with different levels of sensitivity or distribution labels, needed to be handled differently. Most interviewees were not aware of the specifics of governing statutes and regulations, but they were aware of how certain kinds of data or specific labels needed to be handled.

Several interviewees identified DoDI 5230.24, which contains information on DoD distribution statements, as a guide for determining access or release. Interviewees were also well aware of the many DoD CIO–issued directives that encourage data sharing.

Data Flows

Most of the interviewees in the analysis-oriented offices, including the direct support contractors and FFRDCs that assist these offices with analyses, indicated that they accessed the data they needed directly from central repositories.

Interviewee Recommendations

Consistent with the other groups of offices, interviewees in the analysis offices suggested a wide range of solutions to address different aspects of their perceived data-access issues. Again, these recommendations are not derived from RAND and require further analysis before they could be proposed for implementation. Interviewees' recommendations included the following:

- Elevate the problem to higher levels (e.g., OUSD[AT&L]) when acquisition personnel do not get the information needed to do their jobs.
- Persistence: Make a lot of noise (i.e., frequent follow-up requests).
- Find an alternate data source if data barriers cause too much inefficiency.

- Use personal relationships and informal data sharing.
- Articulate the value of a more open access policy in DoD as a whole and in specific communities. Similarly, the benefits of open access and data sharing could be explicitly demonstrated.
- Categorize or "bucket" people according to the level of access they require for their job (e.g., government, FFRDC, UARC, or support contractor). This would represent a new policy. Similarly, associate data-access permissions with the existing security clearance process and have those permissions move with the individual just like his or her clearance.
- Change the incentives for data custodians to encourage granting access and data sharing.
- Establish a coherent policy for which data are legitimately proprietary.
- Include data-access provisions for support contractors in contracts.
- Centralize data handling and coordinate data-access policies for marking data, and make it easier to re-mark data and documents. Establish clear authority for access policy and decisions.
- Streamline the process for moving and using unclassified data from a classified network.

No one in DoD has unfettered access to all data from all sources; information is compartmentalized to avoid large-scale loss and is restricted by need to know. Concerns about information security contribute to the data-sharing problem. Nevertheless, many interviewees noted that there was an opportunity cost for restricting data access and suggested that evaluating the value proposition of investing in data and databases, along with more open access in this context (e.g., a cost-benefit analysis), might help direct needed changes in policy and practice.

Service-Level Interviews

In this section, we focus on the limited number of interviews conducted with service-level acquisition personnel. We spoke with 14 people in five interview sessions. For this study, our main focus was on understanding data sharing from the OSD perspective, but it is unwise to exclude service-level staff, given that much of the data relevant to the other users interviewed originate in the services. Therefore, we spoke with some service-level acquisition personnel to gain a basic understanding of the problem from a service perspective and to collect recommendations from this group.

The service-level personnel with whom we spoke included subject-matter and functional experts, as well as several personnel with a process role or mission. These roles involve executing acquisition programs and sharing data with OSD for decisionmaking and oversight, as needed. We did not speak with anyone in the services who performed crosscutting analysis. As can be expected, our interviewees tended to get the acquisition data they needed to perform their jobs from program offices, prime contractors, or central repositories in the services. They also consulted OSD repositories as secondary sources.

Interviewees described using various types of acquisition data, including cost, technical information, technical parameters, test plans, schedule information, and program management head counts. As with the other offices, it is important to note that we did not interview a large sample of personnel from the services. In terms of labels or markings, PROPIN, FOUO, and pre-decisional were mentioned, with PROPIN data mentioned most often.

Problems Accessing Data

Service-level interviewees agreed with OSD personnel on some types of data-sharing problems. For instance, they agreed that sharing tends to be personality dependent, and some people are more willing to share than others. They also agreed that localized data-sharing "rules" and "practices" are in place, which leads to inconsistent practices across DoD. One example was that need to know was not consistently applied.

There was also agreement on some of the same problems regarding the use of NDAs for accessing proprietary data. Specifically, there is significant variation in the content of NDAs. NDAs can be so restrictive that it takes a lot of effort to get data. Furthermore, companies do not want to create a conflict of interest that compromises existing NDAs.

There was some difference in how both the services and OSD viewed sharing with each other. OSD personnel tended to want quick access to information generated by contractors and the services, but service personnel believed that there needed to be adequate time to scrutinize and prepare raw data to avoid problems with interpretation. In other words, data need to be adjudicated before sharing. The timing of data sharing was also mentioned as an issue. According to interviewees, the services do not have sufficient leeway to perform their function before OSD gets involved. The services also use OSD central repositories as a secondary source for data, and interviewees reported some difficulty getting access to those repositories.

Finally, OSD personnel cite Title 10 authorities that give them generally open access to data from the services; however, service-level interviewees cautioned that there is a formal process or chain of command that should be consulted when it comes to data requests.

Interviewee Recommendations

Some of the recommendations provided by service-level personnel were similar to those offered by OSD personnel. For instance, there was a recommendation for the creation of a repository for the test community's data. Service-level personnel also expressed the need for technical and programmatic data requirements to be included in contracts (e.g., using a contract line item number). In addition, interviewees recommended including an NDA in the initial contract. Creating a standardized NDA form that contractors can sign and that could then be given to each of the prime and subcontractors was another recommendation. There was also a recommendation similar to one provided by OSD interviewees regarding educating leadership about which data are available and making available data visible in one place (e.g., a dashboard). Again, these recommendations are not derived from RAND and require further analysis.

Service-level interviewees were concerned about the number of data requests they received from OSD. They recommended that more time be spent trying to determine which data are truly important in an attempt to reduce the number of data requests. They also recommended coordinating among requestors to see whether the data were already collected.

We received some additional recommendations from these interviewees. For example, they acknowledged that sharing unprocessed data can be problematic from the service perspective, leading to possible misinterpretation. Their recommendation was to allow additional time in the schedule for data processing, which could be set in place by a revision of business rules.

APPENDIX C

Central Repository Case Studies

Over the course of each defense acquisition program's life cycle, a significant amount of data or information is generated on that program that can be used for execution, oversight, and analysis. Major improvements have been made over the past few decades in the ability of organizations to capture and store data generated by these programs in electronic format. One consequence of this improvement is that a significant amount of information is being stored that needs to be accessible for future use. Central repositories are the mechanism for capturing acquisition data, storing them, and releasing them for various uses. This appendix describes seven OSD central repositories that can be used to assist in managing various types of acquisition data.[1] It also provides additional background, details about the benefits and challenges to use, and recommendations for improving acquisition central repositories. The seven repositories are:

- Acquisition Information Repository (AIR)
- Defense Acquisition Management Information Retrieval (DAMIR)
- Defense Automated Cost Information Management System (DACIMS)
- Defense Technical Information Center (DTIC)
- Federal Procurement Data System–Next Generation (FPDS-NG)
- PARCA's Earned Value Management Central Repository (EVM-CR)
- OUSD(AT&L)'s Workforce Data Mart.

The seven repositories all contain various types of acquisition data. More specifically, they contain acquisition information from the information requirements defined by the 2015 DoDI 5000.02. They also include more-detailed cost, budget, earned value, scientific, technical, engineering, contract, and workforce data. Two repositories—DTIC and FPDS-NG—are available to the public. These two repositories are also the largest and most mature of the seven that we reviewed. DTIC traces its roots back to 1945 and FPDS-NG to 1979. The other repositories and parts of DTIC and FPDS-NG have similar access procedures, meaning that users must have a need to know to access sensitive data in the repositories. The typical procedure is to have a "trusted agent" or government sponsor that will vouch for the need for access to certain information. Government employees always have an easier time getting access than contractors because government employees are presumed to have a need to know associated with their official function and are permitted to access proprietary information. The use of a DoD CAC or

[1] We only provide information in this appendix on the unclassified versions of these repositories. For instance, DAMIR has a classified capability, but that is not discussed in this appendix.

PKI is also normally required for access. Users can also get access by having an external certificate authority. Several of these repositories have classified versions, but we address the unclassified versions only. Table C.1 provides more-detailed information on the seven repositories.

Acquisition Information Repository

AIR is the newest of the repositories in this sample. It was deployed in 2012 by USD(AT&L) and is hosted by DTIC.[2] The purpose of AIR is as follows:

> The Department of Defense (DoD) does not routinely store acquisition information required by the current DoD Instruction (DoDI) 5000.02 that the defense enterprise can use to support milestone decision making and analysis. The Office of Enterprise Information and OSD Studies developed the Acquisition Information Repository (AIR) to consolidate acquisition information required by the current DoDI 5000.02. AIR is a searchable document repository that currently stores over 300 unclassified Defense Acquisition Executive–level milestone decision documents related to 105 distinct programs.[3]

AIR has one of the stricter access policies. In addition to using a CAC, users must also access the system from government-furnished equipment. This requirement excludes some potential users who are contractors but may not work in a DoD office (e.g., those who work for FFRDCs). The information in AIR covers all of the 46 information requirements in the interim DoDI 5000.02. Therefore, there are several different markings, including unclassified, FOUO, pre-decisional, and PROPIN. Given that it is a relatively new repository, content is still being populated, and the repository is still building its user list.

Defense Acquisition Management Information Retrieval

DAMIR was stood up in 2004–2005 as

> a DoD initiative that provides enterprise visibility to Acquisition program information. DAMIR streamlines acquisition management and oversight by leveraging web services, authoritative data sources, data collection, and data repository capabilities. DAMIR identifies various data sources that the Acquisition community uses to manage Major Defense Acquisition Programs (MDAP) and Major Automated Information Systems programs and provides a unified web-based interface through which to present that information. DAMIR is the authoritative source for Selected Acquisition Reports (SAR), SAR Baseline, Acquisition Program Baselines, and Assessments . . . and uses web services to obtain and display Defense Acquisition Executive Summary data directly from the Service acquisition databases.[4]

DAMIR is the authoritative source for current and estimated ACAT I and IA program costs. DAMIR is well known for being the main repository of SARs for Congress. It also has begun to add significant capability beyond being a source for document retrieval. Over the past few years, more capability has been added to aid in the analysis of multiple programs across multiple types of acquisition data. DAMIR now has more than 6,000 users from OSD,

---

[2] Kendall, 2012.

[3] Kendall, 2012, p. 1.

[4] Defense Acquisition Management Information Retrieval, homepage, last updated January 13, 2014.

Table C.1
Comparison of OSD Central Repositories

	AIR	DAMIR	DACIMS	DTIC	FPDS-NG	EVM-CR	Workforce Data Mart
Content	Acquisition information required by the interim DoDI 5000.02 (46 information requirements) and an acquisition decision memorandum by USD(AT&L).	SARs, Major Automated Information Systems Annual Report, acquisition program baselines, defense acquisition executive summaries, program objective memoranda, budget estimate submissions, president's budgets, major automated information system annual and quarterly reports, top-level earned value data.	Current and historical cost and software resource data needed to develop independent, substantiated estimates. Includes almost 30,000 Contractor Cost Data Reports, Software Resources Data Reports, and associated documents.	DoD- and government-funded scientific, technical, engineering, and business information available today (public, FOUO, and classified).	Information about federal contracts; allows reporting on federal contracts. Contracts whose estimated value is $3,000 or more. Every modification to that contract, regardless of dollar value, must be reported to FPDS-NG.	Detailed earned value data.	Acquisition workforce data.
Year started	2012	2004–2005	1998	1945	1978 (as FPDS)	2008 in DCARC	2008
Access adjudicator	Office that approves document	ARA	OSD CAPE	Originator of the data	GSA	PARCA	HCI
Repository manager	ARA	ARA	OSD CAPE	DTIC	GSA	PARCA	HCI
Repository host	DTIC	DoD Washington Headquarters Services, the Enterprise Information Technology Services Directorate (EITSD)	OSD CAPE	DTIC	GSA	OSD CAPE	DAU

Table C.1—Continued

	AIR	DAMIR	DACIMS	DTIC	FPDS-NG	EVM-CR	Workforce Data Mart
Access procedures	The AIR help desk verifies a user's identity and need to know with a government sponsor indicated on the DD 2875 form, resulting in appropriate level of access permissions. A DoD CAC/PKI is required for access. Must access via government-furnished computer equipment.	Request account access through organizational trusted agent point of contact. A CAC or PKI is needed, as well as external certificate authority.	User access is granted via either a CAC or trusted certificate. Additional restrictions are applied based on the military branch or service; for example, Army and Air Force users cannot view Navy contracts or submissions. A user's access to information is restricted to programs/contracts within his or her direct area of responsibility. Additionally, a specific submitting contractor, such as Lockheed Martin, cannot review another submitting contractor's, such as Northrop Grumman's, submissions. A CAC or PKI is needed, as well as an external certificate authority and trusted agent point of contact.	Open to the public via the Internet. There is also an access-controlled section where a CAC is required for access. Documents in the access-controlled area require special access is down to the document level (e.g., export controlled).	Need to register as a user of FPDS-NG. Types of users: The system user establishes a link between an agency's contract writing system and FPDS-NG. The government user is able to access, maintain, track, and report contract awards. The public user has access to all reported information in FPDS-NG that is determined ready for release to the general public. The agency system administrator controls and monitors all assigned user types for their respective agency's end users.	User access is granted via either a CAC or trusted certificate. Additional restrictions are applied based on the military branch or service; for example, Army users cannot view Navy contracts or submissions. The user's access to information is restricted to programs/contracts within his or her direct area of responsibility. Additionally, a specific submitting contractor, such as Lockheed Martin, cannot review another submitting contractor's, such as Northrop Grumman's, submissions. A CAC or PKI is needed, as well as an external certificate authority and trusted agent point of contact.	Requires a CAC to access the second level of security, which is username and password based. CAC enabled, U.S. government systems only.
Reasons for restriction	FOUO, pre-decisional, PROPIN (possibly others; document dependent)	FOUO	FOUO, PROPIN, export controlled	No restrictions (publicly releasable); there are some additional restrictions on some of the technical data (e.g., export controlled, classified, FOUO)	No restrictions (publicly releasable)	FOUO, PROPIN, export controlled	FOUO, (personally identifiable information)

Table C.1—Continued

	AIR	DAMIR	DACIMS	DTIC	FPDS-NG	EVM-CR	Workforce Data Mart
Who can access	DoD and defense contractors (see the row on access procedures)	DoD support contractors, defense, FFRDCs, academia, Congress	DoD and defense contractors (required to use NDA process)	Open to the public; the restricted portion open to authorized DoD/U.S. military employees, U.S. government employees, and government contractors and subcontractors	Open to the public; others in government/DoD can access additional information	Government OSD staff access everything; the services can get access to their service program data and can request access to other services' data; contractors can access through NDA process	U.S. government authorized users. Access is restricted to agency workforce managers and functional managers
NDAs required	Yes (contractors)	Yes (contractors)	Yes (contractors)	No (except restricted portions)	No (except restricted portions)	Yes (contractors)	Yes

SOURCES: Interviews conducted for this study; background material available through repositories.

the defense agencies and field activities, FFRDCs, academia, Congress, and the combatant commands.

Defense Automated Cost Information Management System

DACIMS was established in 1998 by OSD CAPE's predecessor, PA&E. DACIMS (along with the EVM-CR) is now hosted in its Cost Assessment Data Enterprise using the DCARC portal.

DACIMS contains around 30,000 contractor cost data reports, software resources data reports, and associated documents.[5] Anyone in DoD can apply for access to DCARC, but some access restrictions may apply. FFRDCs, UARCs, and universities can also apply on an as-needed basis. There are several restrictions on the data in DACIMS, including PROPIN, FOUO, business sensitive, and NOFORN (export controlled). Given that DACIMS contains proprietary data, it is one of the repositories that has been trying to come up with a better way to deal with the many NDAs required to access these data.

Defense Technical Information Center

The DTIC repository is the oldest of the repositories in this case study. DTIC started in 1945:

> The U.S. Army Air Corps, the U.S. Navy and the British Air Ministry establish the Air Documents Research Center in London[,] . . . [which] becomes the Air Document Division of the Intelligence Department of the Headquarters, Air Technical Services, Army Air Force at Wright Field, Ohio.[6]

For nearly 70 years, DTIC has grown into a major central repository for DoD STI. It has multiple functions, including acquiring, storing, retrieving, and disseminating STI. In addition, DTIC hosts more than 100 DoD websites.[7] Unlike some of the other repositories, DTIC is very large and accessible to anyone through an online portal. In addition to this open site, DTIC has a controlled-access repository that requires a CAC or username and password for access. The closed area can be accessed by authorized DoD and U.S. military employees, U.S. government employees, and government contractors and subcontractors.

Federal Procurement Data System–Next Generation

FPDS-NG is a large repository of federal contract information operated by the GSA. As is the case with DTIC, a large portion of the repository is searchable by the general public. FPDS-NG "is also relied upon to create recurring and special reports to the President, Congress, Government Accountability Office, federal executive agencies and the general public."[8] DPAP is the DoD representative for the FPDS-NG. It has been in existence (previously, as FPDS) since October 1, 1978, and currently has more than 60,000 users. Users can extract a wide variety of contract information from the site.

---

[5] Defense Cost and Resource Center, undated(a).

[6] Defense Technical Information Center, "DTIC History," undated.

[7] Pitts, undated, p. 3.

[8] Office of Defense Procurement and Acquisition Policy, 2014.

Earned Value Management Central Repository

EVM-CR is the authoritative source for earned value data in DoD. EVM-CR is a joint effort between USD(AT&L) and CAPE, and it is managed by AT&L/PARCA. According to its website, the repository provides the following:

> Centralized reporting, collection, and distribution for Key Acquisition EVM data
>
> A reliable source of authoritative EVM data and access for OSD, the Services, and the DoD Components
>
> Houses Contract Performance Reports, Contract Funds Status Report, and the Integrated Master Schedules submitted by contractors (and reviewed and approved by Program Management Offices) for ACAT 1C & 1D (MDAP) and ACAT 1A [major automated information system] programs
>
> Approximately 80 ACAT 1A, 1C, and 1D programs and 210 contracts and tasks reporting data.[9]

The repository's user access is based on a CAC or trusted certificate. The repository owner also does not allow the different services to see other service data (e.g., the Navy cannot see Army data). The same is true for contractors. This repository is specifically for earned value, so it has a somewhat smaller user base than some of the broader repositories.

HCI Workforce Data Mart

The Workforce Data Mart has a completely separate data set from the other repositories. The focus is entirely on the acquisition workforce. DAU hosts the repository, but OUSD(AT&L)/HCI directs it. The purpose of the site, according to its guide, is as follows:

> The AT&L DAW Data Mart essentially serves many purposes. It de-conflicts workforce members who may be claimed by multiple components to provide the most accurate AT&L [defense acquisition workforce head] count. The business intelligence tools within the Data Mart provide stakeholders such as the military Services, 4th Estate career managers, [functional integrated process teams], and DoD human capital planners with the capability to run reports to conduct analysis and make strategic decisions regarding the workforce. The data in the Data Mart is populated with data defined in the DoDI 5000.55 plus some additional demographic data elements sourced by the Defense Manpower Data Center.[10]

Access is restricted to government workforce managers, functional managers, and non-government employees whose access is sponsored by DoD, because the repository contains sensitive personnel information.

---

[9] Performance Assessments and Root Cause Analyses, "EVM Central Repository Overview," web page, undated.

[10] Office of the Under Secretary of Defense for Acquisition, Technology and Logistics, Human Capital Initiatives Directorate, and Workforce Management Group, Team B, 2013.

Benefits of OSD Central Repositories

Over the course of our interviews, we discussed the use of central repositories with both users and maintainers of those repositories. One benefit of the central repositories identified by multiple interviewees was that they reduced the burden of fulfilling data requests by the program offices. This is an important benefit; interviewees agreed that program offices, which generate significant amounts of acquisition data, cannot fulfill all requests for data and execute acquisition programs properly. We also heard in our interviews that central repositories are beneficial because they allow analysts to pull information on a variety of topics and programs from one source. This is particularly helpful for analysts who cover multiple programs or multiple types of acquisition data (e.g., cost, schedule, performance, earned value). Each repository has a set of business rules and security policies that identify which groups of people can have access to what information in the repositories. In most but not all cases, repositories are set up to provide immediate access, although in practice access requires obtaining appropriate permissions and therefore does take time. Finally, interviewees observed that central repositories provide a unique historical trail of data that may no longer be accessible due to changes in organizational structure (e.g., program offices that no longer exist).

Problems Identified by Interviewees

Interviewees cited multiple problems with accessing and utilizing central repositories for their work. For example, the various repositories have many scanned documents. Depending on the format, scanned documents are difficult to search (i.e., some are images only that have not been converted to searchable text). Because repositories have grown very large, those that allow queries are more useful than those that do not. Interviewees also stated that not all of their data needs are met by central repositories, which may not have the resources to include everything requested. Prioritizing data needs and capabilities for a central repository will inevitably leave some analysts without all the capabilities that they need. It was mentioned that many central repositories lack OSD-level pre-MDAP information and testing data. In addition, the process of building and populating central repositories takes years. AIR started in 2012 and has been somewhat populated, but there are still more documents that need to be added. Depending on what is provided, it can take a considerable amount of time for personnel to upload all required information—in other words, to institutionalize the process of uploading data to repositories.

Interviewees also agreed on the point that it takes a long time to master using the various central repositories because the software and structure are often very different across databases. Because of this, some interviewees reported that they did not access these repositories regularly, or they relied heavily on staff members with better knowledge of the databases. Some analysts were also unaware of all the available repositories and, consequently, did not know which data could be accessed. Acquisition personnel were aware of the repositories but did not fully use them, or they perceived access and permission as too time-consuming to pursue. In addition, for multiple reasons, there was sometimes a preference for receiving information through working relationships with peers.

Another concern of interviewees was that there was not a centralized or authoritative process for scrubbing and validating all data in a given repository, which may lead to inconsistencies across repositories. In addition, we heard that people or organizations within DoD or contractor organizations are generating data but are not willing to post this information in a repository.

Finally, the use of central repositories as a means of storing, sharing, and analyzing data has increased in DoD over time. The owners of those repositories are faced with a myriad of challenges related to sharing, including integrating information assurance and security policies and procedures, along with business rules, into the architecture of the systems. They also must integrate verification of who can and cannot access which data in the systems. Approving access is not a trivial task, with the thousands of potential users who want access. From the standpoint of those managing repositories, another problem identified during our interviews was that the process of retrofitting systems after the introduction of new security policies or business rules tends to be very cumbersome and time-consuming. One such example involved trying to deal with accounts that become inactive after a certain period of time as dictated by policy. Another was adding a security requirement after the security architecture was defined.

Interviewee Possible Options for Improvement

Interviewees—both those who use repositories and those who manage them—had various options for improvement, but these options were not analyzed by RAND. The first is to formally identify which data in each repository are considered authoritative. There was some confusion about what was considered authoritative during our interviews. Interviewees also observed that there are multiple repositories with different access procedures, business rules, security, and hosts. To analysts seeking and maintaining access, this seemed overly complicated. Standardizing the processes or using one formal entry point for all applications may simplify access procedures. Another possible recommendation is to have a central authority for all of the systems. The central authority would approve access to any of the repositories. It is difficult to obtain access when there are dispersed repositories across OSD. Interviewees also expressed a desire for the central repositories to be easier to access and search.

There are efforts under way to provide integrated views and access from central portals and to improve access to date in these central repositories. ARA's Defense Acquisition Visibility Environment and CAPE's Cost Assessment Data Enterprise are two such portals. Also, many central repositories (both within and outside DoD) provide application programming interfaces to users and portals that allow direct access to data in the repositories. It was beyond the scope of this study to examine these developments in any depth.

APPENDIX D

Technical Data Rights

The authority for DoD and technical data stems from 10 U.S.C. 2302 and 10 U.S.C. 2320. Technical data are defined in 10 U.S.C. 2302 as

> recorded information (regardless of the form or method of the recording) of a scientific or technical nature (including computer software documentation) relating to supplies procured by an agency. Such term does not include computer software or financial, administrative, cost or pricing, or management data or other information incidental to contract administration.[1]

In 10 U.S.C. 2320, rights in technical data are discussed:

> The Secretary of Defense shall prescribe regulations to define the legitimate interest of the United States and of a contractor or subcontractor in technical data pertaining to an item or process. Such regulations shall be included in regulations of the Department of Defense prescribed as part of the Federal Acquisition Regulation. Such regulations may not impair any right of the United States or of any contractor or subcontractor with respect to patents or copyrights or any other right in technical data otherwise established by law. Such regulations also may not impair the right of a contractor or subcontractor to receive from a third party a fee or royalty for the use of technical data pertaining to an item or process developed exclusively at private expense by the contractor or subcontractor, except as otherwise specifically provided by law.[2]

The management of technical data and information is largely an USD(AT&L) responsibility that is implemented through the Federal Acquisition Regulations and the DFARS. Data and information provided to the government by third parties may have markings that have been placed by the third party to identify and protect intellectual property. The type of marking will depend on whether the software or data were developed with government funds. Subpart 227.71 of the DFARS governs the license rights and restrictive markings for technical data developed under a procurement contract, and Subpart 227.72 governs restrictive markings for computer software and software documentation. If the material is not commercial, the DFARS provides for six different protective markings. Each describes the rights to use the information within and outside the government. For noncommercial information or information provided

[1] Legal Information Institute, "10 U.S. Code § 2302—Definitions," Cornell University Law School, undated(a).

[2] Legal Information Institute, "10 U.S. Code § 2320—Rights in Technical Data," Cornell University Law School, undated(b).

to the government outside of a procurement contract, the government's rights will be specified in the contract or whatever legally binding agreement is made.

DoDI 3200.14 defines *scientific and technical information* as follows:

> Communicable knowledge or information resulting from or about the conduct and management of scientific and engineering efforts. [Scientific and technical information] is used by administrators, managers, scientists, and engineers engaged in scientific and technological efforts and is the basic intellectual resource for and result of such efforts. [Scientific and technical information] may be represented in many forms and media. That includes paper, electronic data, audio, photographs, video, drawings, numeric data, textual documents; etc.[3]

Technical data are defined as follows:

> Recorded information related to experimental, developmental, or engineering works that can be used to define an engineering or manufacturing process or to design, procure, produce, support, maintain, operate, repair, or overhaul material. The data may be graphic or pictorial delineations in media, such as drawings or photographs, text in specifications or related performance or design type documents, or computer printouts. Examples of technical data include research and engineering data, engineering drawings, and associated lists, specifications, standards, process sheets, manuals, technical reports, catalog-item identifications, and related information and computer software documentation.[4]

DoD Scientific and Technical Information Program (STIP) policy, led by USD(AT&L), describes a priority for "timely and effective exchange of STI generated by, or needed in, the conduct of DoD [research and engineering] programs." According to DoDI 3200.12,

> The STIP shall permit timely, effective, and efficient conduct and management of DoD research and engineering and studies programs, and eliminate unnecessary duplication of effort and resources by encouraging and expediting the interchange and use of STI. Interchange and use of DoD STI is intended to include the DoD Components, their contractors, other Federal Agencies, their contractors, and the national and international Research & Engineering community.[5]

Sharing of STIP information is determined by distribution statements and classification markings. For studies and analyses performed by contractors, it is up to the DoD sponsoring agency to ensure that the documented research is marked. It is up to the creator of the research or the sponsoring agent to distribute the research. DoDI 3200.14 states that "records shall normally be prepared to allow access and use by DoD contractors and grantees." The instruction also describes the importance of sharing information with contractors: "With the majority of DoD work efforts being performed by contractors and grantees it is essential and in the best interest of the DoD to maximize their access to [STIP] data." It goes on to describe how specific types of information should not be provided to contractors. Planned expenditures or

---

[3] DoDI 3200.14, 2001, p. 14.

[4] DoDI 3200.14, 2001, p. 14.

[5] DoDI 3200.12, 2013.

obligations and "budgetary planning data used to program funds for potential procurement actions that are competition price sensitive" may not be shared with contractors.[6]

Export controls, national security procedures, and protection of proprietary information are enforced for STI that requires such controls and protection.

Technical Data Rights

When the government receives technical data or computer software to which it does not have unlimited license rights, the data or information is required or permitted to have a restrictive marking. Such markings help the government determine what steps are necessary to protect commercial intellectual property (e.g., copyright, patent, trademark, trade secret). DoDI 5230.24 describes the handling of information developed under a procurement contract:

> The DFARS governs the restrictive markings that apply to technical data, documents, or information that is developed or delivered under a DoD procurement contract. Subpart 227.71 of Reference (s) governs the license rights and restrictive markings for technical data, and subpart 227.72 of Reference (s) governs computer software and computer software documentation.

According to DoDI 5230.24,

> The specific format and content of these markings depends on whether the data or software is noncommercial or commercial. In some cases, the applicable regulations specify the precise wording of the restrictive legend, but in many cases the contractor is permitted to use any restrictive legend that appropriately provides notice of the contractor's proprietary interests and accurately characterizes the Government's license rights.[7]

The above instruction seems to contradict how the DFARS explicitly lays out the restrictive markings because it allows the contractor to "use any restrictive legend."

Table D.1 shows the permitted uses within and outside the government for technical data developed or received for noncommercial items. If the government has unlimited rights, there are no restrictions on the sharing of information received. If there are government-purpose rights, the government can use the data for government purposes. Limited rights restrict how the government may use information outside of the government. However, DFARS Subpart 252.227-7013 states that the government may release or disclose data to a covered government support contractor.

If the data in question pertain to a commercial item—meaning they were developed entirely at the expense of the developer—then the permitted uses outside the government and within the government may differ. Table D.2 shows how the government may use data pertaining to commercial items.

Technical documents that have intellectual property marking imposed by a third party, as described in DoDI 5230.24, should be conspicuously and legibly marked with the appropriate legends as prescribed by and controlled pursuant to subpart 252.227-7013 of 203, 227, and 252 of Title 48 of the Code of Federal Regulations.

---

6 DoDI 3200.14, 2001.

7 DoDI 5230.24, 2012, p. 22.

Table D.1
Rights in Noncommercial Computer Software and Technical Data Associated with Noncommercial Items

Rights Category	Applicable to Computer Software or Technical Data	Criteria for Applying Rights Category	Permitted Uses Within Government	Permitted Uses Outside Government
Unlimited	Both	Development exclusive at government expense and any deliverable of certain types, regardless of funding	Unlimited; no restrictions	Unlimited; no restrictions
Government purpose	Both	Development with mixed funding	Unlimited; no restrictions	Only for "government purposes"; no commercial use
Limited	Technical data only	Development exclusively at private expense	Unlimited, except may not be used for manufacture	Emergency repair/ overhaul by foreign government
Restricted	Computer software only	Development exclusively at private expense	Only one computer at a time; minimum backup copies, modification	Emergency repair/ overhaul; certain service/maintenance contracts
Prior government rights	Both	Whenever government has previously acquired rights in the deliverable technical data/computer software	Same as under previous contract	Same as under previous contract
Specifically negotiated license	Both	Mutual agreement of the parties; use whenever the standard rights categories do not meet both parties' needs	As negotiated by the parties; however, must not be less than limited for technical data and must not be less than restricted for computer software	As negotiated by the parties; however, must not be less than limited for technical data and must not be less than restricted for computer software

SOURCE: Richard M. Gray, "Proprietary Information: Everything You Need to Know, but Didn't Know to Ask," briefing, Defense Technical Information Center Conference, March 27, 2007.

NOTE: RAND did not independently verify the contents.

Table D.2
Rights in Commercial Computer Software and Technical Data Pertaining to Commercial Items

Rights Category	Applicable to Technical Data or Computer Software	Criteria for Applying Rights Category	Permitted Uses Within Government	Permitted Uses Outside Government
Unlimited	Technical data only	Any technical data of certain specified types or classes, regardless of commercial status	Unlimited; no restrictions	Unlimited; no restrictions
Standard DFARS 7015 (48 Code of Federal Regulations 252.227-7015)	Technical data only	Default rights category for all technical data pertaining to commercial items except those qualifying for unlimited release, as stated above	Unlimited; except may not be used for manufacture	Only for emergency repair/overhaul
Standard commercial license	Computer software only	Default rights category for all commercial computer software	As specified in the license customarily offered to the public, DoD must negotiate for any specialized needs	As specified in the license customarily offered to the public, DoD must negotiate for any specialized needs
Specifically negotiated license	Both	Mutual agreement of the parties; should be used whenever the standard rights do not meet both parties' needs	As negotiated by the parties; however, by statute, the government cannot accept less than the minimum standard 7015 rights for technical data	As negotiated by the parties; however, by statute, the government cannot accept less than the minimum standard 7015 rights for technical data

SOURCE: Gray, 2007.

NOTE: RAND did not independently verify the contents.

Abbreviations

A&M	Administration and Management
ACAT	acquisition category
AIR	Acquisition Information Repository
ARA	Acquisition Resources and Analysis
ASD(HA)	Assistant Secretary of Defense for Health Affairs
ASD(LA)	Assistant Secretary of Defense for Legislative Affairs
ASD(P&L)	Assistant Secretary of Defense for Production and Logistics
ASD(PA)	Assistant Secretary of Defense for Public Affairs
C3CB	Command, Control, Communication, Cyber and Business Systems
CAC	Common Access Card
CAPE	Cost Assessment and Program Evaluation
CUI	Controlled Unclassified Information
DAB	Defense Acquisition Board
DACIMS	Defense Automated Cost Information Management System
DAMIR	Defense Acquisition Management Information Retrieval
DASD	Deputy Assistant Secretary of Defense
DAU	Defense Acquisition University
DCARC	Defense Cost and Resource Center
DFARS	Defense Federal Acquisition Regulation Supplement
DoD	Department of Defense
DoD CIO	Department of Defense, chief information officer
DoDD	Department of Defense directive
DoDI	Department of Defense instruction

DPAP	Defense Procurement and Acquisition Policy
DT	Developmental Test and Evaluation
DTIC	Defense Technical Information Center
EO	executive order
EVM	Earned Value Management
EVM-CR	Earned Value Management Central Repository
FFRDC	federally funded research and development center
FOIA	Freedom of Information Act
FOUO	For Official Use Only
FPDS-NG	Federal Procurement Data System–Next Generation
GSA	General Services Administration
HCI	Human Capital Initiative
ISOO	Information Security Oversight Office
IT	information technology
MDAP	major defense acquisition program
NDA	nondisclosure agreement
NIPRNet	Non-Secure Internet Protocol Router Network
NOFORN	Not Releasable to Foreign Nationals
OIPT	overarching integrated project team
OPM	Office of Personnel Management
OPSEC	operations security
OSD	Office of the Secretary of Defense
OUSD(AT&L)	Office of the Under Secretary of Defense for Acquisition, Technology and Logistics
PA&E	Programs Analysis and Evaluation
PARCA	Performance Assessments and Root Cause Analyses
PKI	public key infrastructure
PROPIN	proprietary information
SAR	selected acquisition report
SBU	sensitive but unclassified

SE	Systems Engineering
SIPRNet	Secure Internet Protocol Router Network
SSI	Space, Strategic and Intelligence Systems
STI	scientific and technical information
STIP	Scientific and Technical Information Program
TWS	Tactical Warfare Systems
UARC	university-affiliated research center
U.S.C.	United States Code
USD(AT&L)	Under Secretary of Defense for Acquisition, Technology and Logistics
USD(I)	Undersecretary of Defense for Intelligence
USD(P)	Undersecretary of Defense for Policy
USD(P&R)	Undersecretary of Defense for Personnel and Readiness
WHS/HRD	Washington Headquarters Services, Human Resources Directorate

Bibliography

Aftergood, Steve, "What Should Be Classified? Some Guiding Principles," draft, Federation of American Scientists, May 2011.

Carter, Ashton B., "Implementation of the Weapon Systems Acquisition Reform Act of 2009," Directive-Type Memorandum 09-027, Washington, D.C.: Acquisition, Technology and Logistics, Department of Defense, December 4, 2009.

Chairman of the Joint Chiefs of Staff Instruction 6510.01F, *Information Assurance and Support to Computer Network Defense*, February 9, 2011, current as of October 10, 2013.

Controlled Unclassified Information Office, *What Is CIU? Answers to the Most Frequently Asked Questions*, Washington, D.C.: U.S. National Archives and Records Administration, 2011. As of March 29, 2014:
http://www.archives.gov/cui/documents/2011-what-is-cui-bifold-brochure.pdf

Defense Acquisition Management Information Retrieval, homepage, last updated January 13, 2014. As of March 2, 2014:
http://www.acq.osd.mil/damir

Defense Cost and Resource Center, "About DCARC," web page, undated(a). As of March 2, 2014:
http://dcarc.cape.osd.mil/AboutUs.aspx

———, homepage, undated(b). As of March 2, 2014:
http://dcarc.cape.osd.mil

Defense Technical Information Center, "DTIC History," undated. As of March 2, 2014:
http://www.dtic.mil/dtic/pdf/aboutus/dtichistory.pdf

DoDD—*See* U.S. Department of Defense Directive.

DoD Guide—*See* U.S. Department of Defense Guide.

DoDI—*See* U.S. Department of Defense Instruction.

DoD Manual—*See* U.S. Department of Defense Manual.

DoD Regulation—*See* U.S. Department of Defense Regulation.

England, Gordon, "Implementation of the DoD Information Sharing Strategy," memorandum, Washington, D.C.: Department of Defense, August 9, 2007.

Executive Office of the President, *National Strategy for Information Sharing and Safeguarding*, Washington, D.C., December 2012. As of March 29, 2014:
http://www.whitehouse.gov/sites/default/files/docs/2012sharingstrategy_1.pdf

Executive Order 13526, *Classified National Security Information*, Washington, D.C.: The White House, December 29, 2009.

Executive Order 13556, *Controlled Unclassified Information*, Washington, D.C.: The White House, November 4, 2010.

Goitein, Elizabeth, and David. M. Shapiro, *Reducing Overclassification Through Accountability*, New York: Brennan Center for Justice, NYU School of Law, 2011.

Gray, Richard M., "Proprietary Information: Everything You Need to Know, but Didn't Know to Ask," briefing, Defense Technical Information Center Conference, March 27, 2007. As of March 29, 2014: http://www.dtic.mil/dtic/pdf/submit/proprietaryinfogray.pdf

Information Security Oversight Office, Controlled Unclassified Information Notice 2013-01, "Provisional Approval of Proposed CUI Categories and Subcategories," Washington, D.C.: U.S. National Archives and Records Administration, May 22, 2013.

———, *2013 Report to the President*, Washington, D.C.: U.S. National Archives and Records Administration, 2014.

Kendall, Frank, "Approval of Acquisition Information Repository Policy Memorandum," memorandum, Washington, D.C.: Acquisition, Technology and Logistics, Department of Defense, September 4, 2012.

———, "Reliability Analysis, Planning, Tracking, and Reporting," Directive-Type Memorandum 11-003, Washington, D.C.: Acquisition, Technology and Logistics, Department of Defense, January 16, 2013.

Legal Information Institute, "10 U.S. Code § 2302—Definitions," Cornell University Law School, undated(a). As of November 3, 2014:
http://www.law.cornell.edu/uscode/text/10/2302

———, "10 U.S. Code § 2320—Rights in Technical Data," Cornell University Law School, undated(b). As of November 3, 2014:
http://www.law.cornell.edu/uscode/text/10/2320

Libicki, Martin C., Brian A. Jackson, David R. Frelinger, Beth E. Lachman, Cesse Cameron Ip, and Nidhi Kalra, *What Should Be Classified? A Framework with Application to the Global Force Management Data Initiative*, Santa Monica, Calif.: RAND Corporation, MG-989-JS, 2010. As of October 02, 2014: http://www.rand.org/pubs/monographs/MG989.html

National Commission on Terrorist Attacks upon the United States, *The 9/11 Commission Report: Final Report of the National Commission on Terrorist Attacks upon the United States*, Washington, D.C.: U.S. Government Printing Office, July 22, 2004.

National Security Archive, "FOIA Legislative History," web page, undated. As of March 29, 2014: http://www2.gwu.edu/~nsarchiv/nsa/foialeghistory/legistfoia.htm

Office of Defense Procurement and Acquisition Policy, "Federal Procurement Data System–Next Generation (FPDS-NG)," web page, last updated February 27, 2014. As of March 2, 2014:
http://www.acq.osd.mil/dpap/pdi/eb/federal_procurement_data_system_-_next_generation_fpds-ng.html

Office of the Assistant Secretary of Defense for Acquisition, Performance Assessments and Root Cause Analysis Directorate, *Earned Value Management Central Repository (EVM-CR) Data Quality Dashboard User Guide*, undated.

Office of the Under Secretary of Defense for Acquisition, Technology and Logistics; Human Capital Initiatives Directorate; and Workforce Management Group, Team B, *Defense Acquisition Workforce Data and Information Consumer Guide*, Version 1.1, April 11, 2013.

Performance Assessment and Root Cause Analyses, "EVM Central Repository Overview," web page, undated. As of March 2, 2014:
http://dcarc.cape.osd.mil/EVM/EVMOverview.aspx

Pitts, Shari, "DTIC Overview," briefing, Defense Technical Information Center, undated.

Public Interest Declassification Board, *Transforming Security Classification System: Report to the President from the Public Interest Declassification Board*, Washington, D.C., November 2012.

Rhodes, Michael L., "DoD Internal Information Collections," Directive-Type Memorandum 12-004, Washington, D.C., October 23, 2012.

Uniform Law Commission, "Legislative Fact Sheet: Trade Secrets Act," web page, 2014. As of March 29, 2014:
http://www.uniformlaws.org/LegislativeFactSheet.aspx?title=Trade%20Secrets%20Act

United States Code, Title 5, Section 552, Public information; Agency Rules, Opinions, Orders, Records, and Proceedings, January 16, 2014.

———, Title 10, Section 129d, Disclosure to Litigation Support Contractors, January 16, 2014.

———, Title 10, Section 131, Office of the Secretary of Defense, January 7, 2011.

———, Title 10, Section 1761, Management Information System, January 16, 2014.

———, Title 18, Section 1832, Theft of Trade Secrets, January 3, 2012.

———, Title 18, Section 1905, Disclosure of Confidential Information Generally, January 16, 2014.

U.S. Congress, 111th Cong., National Defense Authorization Act for Fiscal Year 2010, Washington, D.C., H.R. 2647, Public Law 111–84, October 28, 2009.

U.S. Department of Defense, "Office of the Secretary of Defense," web page, undated. As of March 29, 2014: http://www.defense.gov/osd

U.S. Department of Defense Administrative Instruction 15, *OSD Records and Information Management Program*, May 3, 2013.

U.S. Department of Defense Administrative Instruction 56, *Management of Information Technology (IT) Enterprise Resources and Services for OSD, Washington Headquarters Services (WHS), and Pentagon Force Protection Agency (PFPA)*, April 29, 2013.

U.S. Department of Defense Administrative Instruction 101, *Personnel and Data Management Information Reporting Policies and Procedures for Implementation of the Defense Acquisition Workforce Improvement Act (DAWIA)*, July 20, 2012.

U.S. Department of Defense, Defense Federal Acquisition Regulation Supplement, Government Support Contractor Access to Technical Data (DFARS 2009-D031), *Federal Register*, Final Rule, May 22, 2013. As of March 29, 2014:
https://www.federalregister.gov/articles/2013/05/22/2013-12055/defense-federal-acquisition-regulation-supplement-government-support-contractor-access-to-technical

U.S. Department of Defense, Defense Federal Acquisition Regulation Supplement, Subpart 227.71, "Rights in Technical Data," Defense Procurement and Acquisition Policy, revised February 28, 2014. As of March 28, 2014:
http://www.acq.osd.mil/dpap/dars/dfars/html/current/227_71.htm

U.S. Department of Defense and Defense Information Systems Agency, *Access Control in Support of Information Systems Security Technical Implementation Guide, version 2, release 1*, October 17, 2007.

U.S. Department of Defense Directive 3600.01, *Information Operations (IO)*, May 2, 2013.

U.S. Department of Defense Directive 4630.05, *Interoperability and Supportability of Information Technology (IT) and National Security Systems (NSS)*, May 5, 2004, current as of April 23, 2007.

U.S. Department of Defense Directive 5000.01, *The Defense Acquisition System*, May 12, 2003, current as of November 20, 2007.

U.S. Department of Defense Directive 5015.2, *DoD Records Management Program*, March 6, 2000.

U.S. Department of Defense Directive 5105.73, *Defense Technical Information Center (DTIC)*, May 2, 2013.

U.S. Department of Defense Directive 5134.01, *Under Secretary of Defense for Acquisition, Technology, and Logistics (USD[AT&L])*, December 9, 2005, incorporating change 1, April 1, 2008.

U.S. Department of Defense Directive 5134.13, *Deputy Under Secretary of Defense for Acquisition and Technology (DUSD[A&T])*, October 5, 2005.

U.S. Department of Defense Directive 5134.14, *Principal Deputy Under Secretary of Defense for Acquisition, Technology, and Logistics (PDUSD[AT&L])*, December 29, 2010, incorporating change 1, May 4, 2011.

U.S. Department of Defense Directive 5143.01, *Under Secretary of Defense for Intelligence (USD[I])*, November 23, 2005.

U.S. Department of Defense Directive 5144.02, *DoD Chief Information Officer (DoD CIO)*, April 22, 2013.

U.S. Department of Defense Directive 5200.27, *Acquisition of Information Concerning Persons and Organizations Not Affiliated with the Department of Defense*, January 7, 1980.

U.S. Department of Defense Directive 5230.09, *Clearance of DoD Information for Public Release*, August 22, 2008, certified current through August 22, 2015.

U.S. Department of Defense Directive 5230.11, *Disclosure of Classified Military Information to Foreign Governments and International Organization*, June 16, 1992.

U.S. Department of Defense Directive 5230.25, *Withholding of Unclassified Technical Data from Public Disclosure*, November 6, 1984, incorporating change 1, August 18, 1995.

U.S. Department of Defense Directive 5400.11, *DoD Privacy Program*, May 8, 2007, incorporating change 1, September 1, 2011.

U.S. Department of Defense Directive 8000.01, *Management of the Department of Defense Information Enterprise*, February 10, 2009.

U.S. Department of Defense Directive 8500.01E, *Information Assurance (IA)*, October 24, 2002, certified current as of April 24, 2007.

U.S. Department of Defense Guide 8320.02-G, *Guidance for Implementing Net-Centric Data Sharing*, April 12, 2006.

U.S. Department of Defense Instruction 1312.01, *Department of Defense Occupational Information Collection and Reporting*, January 28, 2013.

U.S. Department of Defense Instruction 1336.08, *Military Human Resource Records Life Cycle Management*, November 13, 2009.

U.S. Department of Defense Instruction 2015.4, *Defense Research, Development, Test and Evaluation (RDT&E) Information Exchange Program (IEP)*, February 7, 2002.

U.S. Department of Defense Instruction 2040.02, *International Transfers of Technology, Articles, and Services*, July 10, 2008, revised March 27, 2014.

U.S. Department of Defense Instruction 3025.19, *Procedures for Sharing Information with and Providing Support to the U.S. Secret Service (USSS), Department of Homeland Security (DHS)*, November 29, 2011.

U.S. Department of Defense Instruction 3200.12, *DoD Scientific and Technical Information Program (STIP)*, August 22, 2013.

U.S. Department of Defense Instruction 3200.14, *Principles and Operational Parameters of the DoD Scientific and Technical Information Program*, May 13, 1997, administrative reissuance incorporating through change 3, June 28, 2001.

U.S. Department of Defense Instruction 4630.8, *Procedures for Interoperability and Supportability of Information Technology (IT) and National Security Systems (NSS)*, June 30, 2004.

U.S. Department of Defense Instruction 5000.2, *Operation of the Defense Acquisition System*, January 7, 2015.

U.S. Department of Defense Instruction 5000.35, *Defense Acquisition Regulations (DAR) System*, October 21, 2008.

U.S. Department of Defense Instruction 5000.55, *Reporting Management Information on DoD Military and Civilian Acquisition Personnel and Positions*, November 1, 1991.

U.S. Department of Defense Instruction 5030.59, *National Geospatial-Intelligence Agency (NGA) Limited Distribution Geospatial Intelligence*, December 7, 2006.

U.S. Department of Defense Instruction 5040.02, *Visual Information (VI)*, October 27, 2011.

U.S. Department of Defense Instruction 5200.01, *DoD Information Security Program and Protection of Sensitive Compartmented Information*, October 9, 2008, incorporating change 1, June 13, 2011.

U.S. Department of Defense Instruction 5200.39, *Critical Program Information (CPI) Protection Within the Department of Defense*, July 16, 2008, incorporating change 1, December 28, 2010.

U.S. Department of Defense Instruction 5200.44, *Protection of Mission Critical Functions to Achieve Trusted Systems and Networks (TSN)*, November 5, 2012.

U.S. Department of Defense Instruction 5205.08, *Access to Classified Cryptographic Information*, November 8, 2007.

U.S. Department of Defense Instruction 5210.02, *Access to and Dissemination of Restricted Data and Formerly Restricted Data*, June 3, 2011.

U.S. Department of Defense Instruction 5210.83, *DoD Unclassified Controlled Nuclear Information (UCNI)*, July 12, 2012.

U.S. Department of Defense Instruction 5230.24, *Distribution Statements on Technical Documents*, August 23, 2012.

U.S. Department of Defense Instruction 5230.27, *Presentation of DoD-Related Scientific and Technical Papers at Meetings*, October 6, 1987.

U.S. Department of Defense Instruction 5230.29, *Security and Policy Review of DoD Information for Public Release*, January 8, 2009, revised August 13, 2014.

U.S. Department of Defense Instruction 5400.04, *Provision of Information to Congress*, March 17, 2009.

U.S. Department of Defense Instruction 5535.11, *Availability of Samples, Drawings, Information, Equipment, Materials, and Certain Services to Non-DoD Persons and Entities*, March 19, 2012.

U.S. Department of Defense Instruction 6040.40, *Military Health System Data Quality Management Control Procedures*, November 26, 2002.

U.S. Department of Defense Instruction 8110.1, *Multinational Information Sharing Networks Implementation*, February 6, 2004.

U.S. Department of Defense Instruction 8260.2, *Implementation of Data Collection, Development, and Management for Strategic Analyses*, January 21, 2003.

U.S. Department of Defense Instruction 8320.02, *Sharing Data, Information, and Information Technology (IT) Services in the Department of Defense*, August 5, 2013.

U.S. Department of Defense Instruction 8330.01, *Interoperability of Information Technology (IT), Including National Security Systems (NSS)*, May 21, 2014.

U.S. Department of Defense Instruction 8500.01, *Cybersecurity*, March 14, 2014.

U.S. Department of Defense Instruction 8500.2, *Information Assurance (IA) Implementation*, February 6, 2003.

U.S. Department of Defense Instruction 8510.01, *DoD Information Assurance Certification and Accreditation Process (DIACAP)*, November 28, 2007.

U.S. Department of Defense Instruction 8510.01, *Risk Management Framework (RMF) for DoD Information Technology (IT)*, March 12, 2014.

U.S. Department of Defense Instruction 8520.02, *Public Key Infrastructure (PKI) and Public Key (PK) Enabling*, May 24, 2011.

U.S. Department of Defense Instruction 8520.03, *Identity Authentication for Information Systems*, May 13, 2011.

U.S. Department of Defense Instruction 8550.01, *DoD Internet Services and Internet-Based Capabilities*, September 11, 2012.

U.S. Department of Defense Instruction 8580.1, *Information Assurance (IA) in the Defense Acquisition System*, July 9, 2004.

U.S. Department of Defense Instruction 8582.01, *Security of Unclassified DoD Information on Non-DoD Information Systems*, June 6, 2012.

U.S. Department of Defense Instruction 8910.01, *Information Collection and Reporting*, March 6, 2007, incorporating change 1, May 19, 2014.

U.S. Department of Defense Interim Instruction 5000.02, *Operation of the Defense Acquisition System*, November 25, 2013.

U.S. Department of Defense Manual 5000.04-M-1, *Cost and Software Data Reporting (CSDR) Manual*, November 4, 2011.

U.S. Department of Defense Manual 5010.12-M, *Procedures for the Acquisition and Management of Technical Data*, May 14, 1993.

U.S. Department of Defense Manual 5105.21, Vol. 1, *Sensitive Compartmented Information (SCI) Administrative Security Manual: Administration of Information and Information Systems Security*, October 19, 2012.

U.S. Department of Defense Manual 5105.21, Vol. 2, *Sensitive Compartmented Information (SCI) Administrative Security Manual: Administration of Physical Security, Visitor Control, and Technical Security*, October 19, 2012.

U.S. Department of Defense Manual 5105.21, Vol. 3, *Sensitive Compartmented Information (SCI) Administrative Security Manual: Administration of Personnel Security, Industrial Security, and Special Activities*, October 19, 2012.

U.S. Department of Defense Manual 5200.01, Vol. 1, *DoD Information Security Program: Overview, Classification, and Declassification*, February 24, 2012.

U.S. Department of Defense Manual 5200.01, Vol. 2, *DoD Information Security Program: Marking of Classified Information*, February 24, 2012, incorporating change 2, March 19, 2013.

U.S. Department of Defense Manual 5200.01, Vol. 3, *DoD Information Security Program: Protection of Classified Information*, February 24, 2012, incorporating change 2, March 19, 2013.

U.S. Department of Defense Manual 5200.01, Vol. 4, *DoD Information Security Program: Controlled Unclassified Information (CUI)*, February 24, 2012.

U.S. Department of Defense Manual 8400.01-M, *Procedures for Ensuring the Accessibility of Electronic and Information Technology (E&IT) Procured by DoD Organizations*, June 3, 2011.

U.S. Department of Defense Manual 8910.01, Vol. 1, *DoD Information Collections Manual: Procedures for DoD Internal Information Collections*, June 30, 2014.

U.S. Department of Defense Manual 8910.1-M, *Department of Defense Procedures for Management of Information Requirements*, June 30, 1998.

U.S. Department of Defense Regulation 5200.1-R, *Information Security Program*, January 1997.

U.S. Department of Defense Regulation 5400.7-R, *DoD Freedom of Information Act Program*, September 1998.

U.S. Department of Justice, "FOIA Resources," web page, undated. As of March 29, 2014:
http://www.justice.gov/oip/foia-resources.html

U.S. Government Accountability Office, *Testimony Before the Subcommittee on National Security, Emerging Threats, and International Relations, Committee on Government Reform, House of Representatives, Managing Sensitive Information: DOE and DOD Could Improve Their Policies and Oversight*, Washington, D.C., GAO-06-531T, March 14, 2006.

———, *Defense Acquisitions: Assessments of Selected Weapon Programs*, Washington, D.C., GAO-14-340SP, March 2014.

U.S. National Archives and Records Administration, "Controlled Unclassified Information," web page, undated. As of March 29, 2014:
http://www.archives.gov/cui/registry/category-list.html

White House, Office of the Press Secretary, "Designation and Sharing of Controlled Unclassified Information (CUI)," memorandum, May 7, 2008.